TEACHING FOR ECOJUSTICE

"This is a very timely book! The growing field of EcoJustice Education needs a book that helps teachers and teacher educators translate complex analytic material into classroom practices and lessons. We have been waiting for this book!"

Rebecca Martusewicz, Eastern Michigan University, USA

"This book translates academic and theoretical works on EcoJustice into accessible curricular materials designed to equip students to reflect critically on cultural roots of the environmental crisis unfolding on the planet. An important strength of the sample lesson plans is that they assume agency on the part of the teacher-reader to adjust these learning activities for student needs in specific contexts. Dr. Turner conveys a sense of respect for the teacher-reader's professional judgment."

Teresa Shume, Minnesota State University Moorhead, USA

"This lively, relevant, and timely text fills a need for practical work in EcoJustice Education."

Audrey M. Dentith, Lesley University, USA

Teaching for EcoJustice is a unique resource for exploring the social roots of environmental problems in humanities-based educational settings and a curriculum guidebook for putting EcoJustice Education into practice. It provides model curriculum materials that apply the principles of EcoJustice Education, giving pre- and in-service teachers the opportunity to review examples of specific secondary and post-secondary classroom assignments, lessons, discussion prompts, and strategies that encourage students to think critically about how modern problems of sustainability and environmental destruction have developed, what their root causes are, and how these problems can be addressed. The author describes instructional methods she uses when teaching each lesson and shares insights from evaluations of the materials in her classroom and by other teachers. Interspersed between lessons is commentary about the rationale behind the materials and observations about their effect on students.

Rita J. Turner, Ph.D., is a lecturer in American Studies and Media and Communication Studies at the University of Maryland, Baltimore County, USA.

TEACHING FOR ECOJUSTICE

Curriculum and Lessons for Secondary and College Classrooms

Rita J. Turner

Routledge
Taylor & Francis Group

NEW YORK AND LONDON

First published 2015
by Routledge
711 Third Avenue, New York, NY 10017

and by Routledge
2 Park Square, Milton Park, Abingdon, Oxon OX14 4RN

Routledge is an imprint of the Taylor & Francis Group, an informa business

© 2015 Taylor & Francis

Library of Congress Cataloging in Publication Data
Turner, Rita J.
Teaching for ecojustice : curriculum and lessons for secondary and college
classrooms / by Rita J. Turner.
pages cm
Includes bibliographical references and index.
1. Environmental education. 2. Human ecology—Study and teaching
(Secondary) 3. Human ecology—Study and teaching (Higher) I. Title.
GF26.T87 2015
363.70071—dc23
2014048312

ISBN: 978-1-138-83291-6 (hbk)
ISBN: 978-1-138-83292-3 (pbk)
ISBN: 978-1-315-73568-9 (ebk)

Typeset in Bembo
by Swales and Willis Ltd, Exeter, Devon, UK

Printed and bound in the United States of America by Publishers Graphics,
LLC on sustainably sourced paper.

For Boogie, my heart and my teacher.

And with endless love and gratitude to Tom Turner, the best dad in the world; to Ryan Donnelly, whose constant love and partnership I cherish and rely on; and to my home and yard for nourishing and sheltering me.

CONTENTS

PREFACE

Any book published today that emerges from even a partial awareness of the state of the world will likely begin in an increasingly familiar way. It will establish that we face enormous social and environmental challenges in the world. It will describe some of the looming crises to which humankind must respond, including climate change, social inequality, food and water security, and the threat of mass extinctions. Having set the stage, it will offer its own contributions to the questions of how these overwhelming problems came to be and how they can be changed for the better. I have read many books that begin this way, because they must; this is the reality of our world.

This book, too, emerges from an awareness of these dire circumstances, and from a desire to contribute to their resolution. It follows on from important works that have offered powerful insights into the questions of how the world came to be as it is and how it can be remade for greater justice and sustainability.

In the realm of education, these works arise from the emerging fields of EcoJustice Education, ecopedagogy, and critical ecoliteracy. Scholars in these fields are thinking seriously about how our practice of teaching can respond to planetary crisis and can help humanity to develop new relationships to each other, other living beings, and the surrounding earth.

The insights offered by this scholarship are compelling. Scholars in EcoJustice – as well as important theorists in environmental philosophy, cultural studies, ecofeminism, critical discourse studies, and beyond – emphasize that our actions spring from and are influenced by cultural frameworks of belief. They contend that our belief systems shape what we think about our role in the world, the role of other beings, and how we should live on this planet. And they argue that we can learn to question our ingrained belief systems, recognize their implications, and rethink them in order to achieve more positive and sustainable behavior (Abram

1997; Bowers 2001; Kahn 2009; Martusewicz, Edmundson, and Lupinacci 2014; Peterson 2009; Smith 2001; Turner 2011; Warren 2000). These ideas are vital to helping us understand why we live in the world the way we do and how we could make a better future, and they should deeply influence our teaching and our thinking. But for classroom teachers and teachers-in-training learning about these important perspectives, the next challenge is to apply them to curriculum design and classroom instruction. I hope the lessons in this book will give teachers a model of how this can be done.

Teaching for EcoJustice: Curriculum and Lessons for Secondary and College Classrooms demonstrates how a teacher might directly apply the insights of EcoJustice Education and related fields of scholarship in the classroom. It is a curriculum guidebook for teachers-in-training and in-service teachers who want to learn strategies for exploring the social roots of environmental problems, particularly in humanities-based educational settings including English/language arts and social studies classrooms. This book offers curriculum materials that apply the principles of EcoJustice Education, giving those in teacher education programs and classroom teachers the opportunity to review examples of specific secondary and post-secondary classroom assignments, lessons, discussion prompts, and strategies that encourage students to think critically about how modern problems of sustainability and environmental destruction have developed, their root causes, and how they can be addressed. All of the lessons in this book come from my own classroom, where I've designed and tested them for several years. I'm excited to share them here, to help teachers put the vital goals of EcoJustice Education into day-to-day practice.

We know the problems. Thanks to the work of great thinkers in education and other fields, the stage is set. Now it's time for teachers and students to get out on that stage and dance.

What's in This Book

This book contains eight units that address different topics of study I believe are important to explore in the EcoJustice classroom, as well as an Introduction and a Conclusion chapter. Here is a brief overview of the chapters and units:

- Introduction. In this opening chapter I describe my definitions of EcoJustice Education and why I believe it is so important to incorporate it into classrooms. I explain why studying culture, discourse, and patterns of thought is necessary in order to create a more environmentally and socially just world, and I discuss how EcoJustice lessons can be incorporated into the goals of a traditional classroom. I then discuss pedagogical considerations and explain the details of what you will find in each lesson.
- Unit 1: Nature and the Self. The lessons in this unit ask students to think about their relationship with the natural world and with the nonhuman beings that co-inhabit this world with them, and to consider both what

impact the natural world has on them and what impact they have on the natural world and on other beings.

- Unit 2: Language, Media, and Worldviews. By exploring diverse worldviews and studying the root metaphors we use to reason about the natural world, this unit helps students understand the patterns of thought that have led modern society toward destructive interactions with the larger world, and explore alternative patterns of thought that do not support oppression or exploitation.

- Unit 3: Place. This unit is designed to help students explore the significance of place in their lives and in the lives of others. Lessons ask students to reflect on their own memories of places that are important to them and think deeply about how "place" and "home" are experienced today for humans and for other species.

- Unit 4: Food. This unit focuses on our modern food system and its social and environmental impacts. This includes considering how food is grown, how those who grow, ship, and prepare food are treated, how modern processed food affects health, who does and does not have easy access to healthy food, and how other species are treated, including how nonhuman animals are treated in factory farms.

- Unit 5: Stuff – Production, Consumption, and Waste. The lessons in this unit focus on how consumer goods are made, how and why we consume these goods as we do, and what happens to them once we dispose of them. Students will learn what goes into making our "stuff" and the environmental and social impacts of this production and consumption.

- Unit 6: Environmental Attitudes and Behaviors in U.S.-American History. This unit explores how attitudes toward the environment have changed over time in the U.S. Students study opinions about the land and humankind's role within it from historical writings over the early history of the U.S. into the twentieth century.

- Unit 7: Ethics and Environmental Justice. This unit tackles questions of environmental ethics, animal rights, and environmental justice. Assignments ask students to compare the ethical arguments of different authors, reflect on the motivations behind moral and legal decision-making, and extrapolate how society would need to change in order to align with different ethical ideals.

- Unit 8: Imagining Possible Futures. The lessons in this unit ask students to respond to disparate visions of the future and to explore proposed changes to our modern ways of living. Assignments engage students in creative imagining of new future paths.

- Conclusion. In the concluding chapter I reflect on what I hope students and teachers will get from the lessons in this book. I suggest that educators should encourage empathy, dialogue, questioning of dominant assumptions, analysis of culture, and ethical reflection in their classrooms, and should create a space for students to express their desire for a better world and for them to work together to imagine creative strategies that might help achieve that better world.

Within each of the eight units in this book are individual lessons; for each one I describe materials, assignments, and procedures for teaching the lesson. Each lesson uses varied materials, from poetry and essays to fictional prose, documentary film, artwork, newspaper articles, and videos. Included in the lessons are writing assignments and other activities I assign to my students, as well as examples of writing that my own students have done in response to these assignments.

At the end of each unit, you will find a section titled "Effect of the Lessons." In these sections I discuss key skills or qualities that the lessons in that particular unit (and in the curriculum in general) should help to cultivate in students. I highlight examples of my own students demonstrating their engagement with the skill or quality that I'm discussing, and I talk about why that particular skill is so important.

There is also an online resource for this book that contains additional materials for teachers and teachers-in-training, including text of some of the readings assigned in the lessons, relevant websites, links to standardized learning objectives met by each lesson, and additional activities. You can access the online resource at: www.routledge.com/9781138832923.

How to Use This Book

For teachers-in-training, this book can serve as a model of instructional design in EcoJustice, helping you to explore how to design lessons that encourage critical thinking about important social and environmental issues. You may choose to read this book in linear fashion from start to finish, or you may choose, after reading the Introduction, to focus only on certain units. I hope these lessons prove to be fruitful examples of EcoJustice Education in the classroom as you build your own educational praxis. Each model lesson offers an example of curriculum design that combines skills and content essential to EcoJustice Education with skills in language use, information use, and critical multiliteracies that are already an essential part of English/language-arts, social studies, and other humanities-oriented classrooms.

For in-service teachers, if you teach humanities-based courses at the high school or college level such as English, social studies, literature, writing, ethics, history, civics, cultural studies, or media studies, the lessons in this book are ready to be put to use in your classroom. You can either teach them exactly as described, following my procedures and reproducing the sample assignments to give to your students, or you can modify them to better match the skill level and theme of your particular students and course. You may decide to use as many or as few lessons as you like in your own classroom, either selecting individual lessons, teaching entire units, or teaching the entire curriculum from Unit 1 through Unit 8. If you do teach multiple lessons or multiple units, I encourage you to teach them in the order they are presented. Using the lessons in order isn't essential, but I've designed and taught these materials so that the ideas and skills will build on one another from one lesson to the next, so I recommend it when possible. I teach these lessons as a contiguous curriculum and I'll be thrilled to see others do the same, but I'm also eager to see teachers use their enormous wealth of innovation, creativity, and skill to modify these lessons to suit their own students' needs, adding or cutting materials, tying the lessons into other topics and other assignments, linking them to local issues, and more.

The lessons can be applied to students with a range of skill levels at the high school, community college, and undergraduate college levels, and I encourage teachers to use their judgment as to which readings and assignments best suit the skill level of their students.

I also encourage you to pair this book with texts that address the theory of EcoJustice Education, ecopedagogy, and critical ecoliteracy in more depth, as these provide important background about the ideas and reasons behind teaching materials like the ones you'll find here. *EcoJustice Education: Toward Diverse, Democratic, and Sustainable Communities* by Martusewicz, Edmundson, and Lupinacci would serve as an excellent companion to this book, and I hope teachers, prospective teachers, and teacher educators will read both.

References

Abram, David. 1997. *The Spell of the Sensuous: Perception and Language in a More-Than-Human World.* 1st Vintage Books Ed. New York: Vintage.

Bowers, C. A. 2001. *Educating for Eco-Justice and Community.* Athens: University of Georgia Press.

Kahn, Richard. 2009. "Towards Ecopedagogy: Weaving a Broad-Based Pedagogy of Liberation for Animals, Nature, and the Oppressed People of the Earth." In *The Critical Pedagogy Reader,* edited by Antonia Darder, Marta P. Baltodano, and Rodolfo Torres, 2nd ed., 522–40. New York: Routledge.

Martusewicz, Rebecca A., Jeff Edmundson, and John Lupinacci. 2014. *EcoJustice Education: Toward Diverse, Democratic, and Sustainable Communities.* 2nd ed. New York: Routledge.

Peterson, Anna Lisa. 2009. *Everyday Ethics and Social Change: The Education of Desire.* New York: Columbia University Press.

Smith, Mick. 2001. *An Ethics of Place: Radical Ecology, Postmodernity, and Social Theory.* Albany, NY: State University of New York Press.

Turner, Rita. 2011. "Critical Ecoliteracy: An Interdisciplinary Secondary and Postsecondary Humanities Curriculum to Cultivate Environmental Consciousness." Baltimore, MD: University of Maryland, Baltimore County.

Warren, Karen. 2000. *Ecofeminist Philosophy: A Western Perspective on What It Is and Why It Matters.* Lanham, MD: Rowman & Littlefield.

ACKNOWLEDGMENTS

I wish to express profound thanks to my students, for going on this journey with me, for being such sources of inspiration and insight along the way, and for the openness of your minds and your hearts. Also a special thanks to those students who granted permission for me to include their work in this book: Yiannis Balanos, Chelsie Bateman, Mallory Brooks, Danny Clemens, Will Fejes, Heather Harshbarger, Christina Malliakos, Malarie Novotny, Michelle Ott, Dwarka Pazavelil, Rebecca Postowski, Jacob Rosenborough, Ashley Sweet, Jennie Williams, and two students who wish to remain anonymous. Your words are extremely meaningful to me, and I'm honored to have the opportunity to share them with others.

My deepest thanks as well to my friends and colleagues who have supported this project. Thanks to Jeanine Williams for being a sounding-board, writing companion, and friend. To Christine Mallinson, for her valuable feedback on chapters of this manuscript and her consistent guidance and encouragement. To Johnny Lupinacci for his input on chapters of this manuscript and his support and excitement. And to Rebecca Martusewicz, Richard Kahn, Bev Bickel, Joby Taylor, Ed Orser, and Mary Rivkin for helping this project along at various points in indispensible ways. I'm so grateful to each of you.

I'd also like to express my enormous gratitude to Naomi Silverman for her help and support, and to the readers who reviewed my prospectus for this book: Rebecca Martusewicz of Eastern Michigan University, Audrey M. Dentith of Lesley University, and Teresa Shume of Minnesota State University. I was honored by the thoughtful, enthusiastic, and helpful feedback you provided.

INTRODUCTION

Why Teach for EcoJustice?

The purpose of this book is to help us think about our world and what we want it to be. It's about recognizing injustices and getting to the bottom of how and why they happen. It's about knowing that what we think and believe shapes what we do, and that our beliefs don't come from nowhere, there are forces shaping them as well. And it's about making active choices about how we want our world to be in the future.

When we look at the world today, if we let ourselves look closely, we'll see many things we want to be different. We'll see a heartbreaking scope of destruction and cruelty – humans, other animals, plants, and whole ecosystems suffering and dying as they are exploited, uprooted, poisoned, and abused. If we look at the possibilities for our future, if we read the studies or simply watch the changes around us, we'll see even darker trends ahead. This is not what I want my world to be like, and I'm fairly confident this isn't what anyone else wants their world to be like either. We want joy, and justice, and abundance. We want to flourish, and we want others to flourish. We want a healthy community and a healthy ecosystem. If we could choose between living lives that deplete our world or living lives that nourish it, we would choose to nourish. If we could choose between the daunting future we see on the horizon or a future of mutual well-being, we would choose mutual well-being. So how do we enact that choice, how do we bring that future into being?

Teachers see the future in front of them every day. We watch future leaders, future parents, future caretakers develop before our eyes. We know that what students learn, what they believe, and how they see themselves and their world can shape who they become as people. And we know that who our students become as people can shape the world.

I'm a teacher because I'm not satisfied with the world as it is. I'm a teacher because I want to see people shape the world for the better. I want to see a future that is more just, healthy, and happy than the world I see around me today. A future in which humans aren't destroying the ecosystems we depend on for survival. In which my well-being and comfort do not rely on the suffering of others, either humans or nonhumans. In which being a member of an ecosystem means contributing positively to the lives of others, not simply moderating how much I can take. In which the family of plants and animals I have grown up with in my yard and town and bioregion can continue to thrive, and are not driven extinct or forced from my region due to climate change or habitat loss.

I want to see the flourishing of my neighborhood, my city, my watershed, my ecosystem because they are part of me and I of them. I want to see them flourish because I love them. In my life I have had the immeasurable privilege of learning about love not only from my human relatives and friends, but also from other species and from the land. I have experienced the gift of spending 19 years with a dog companion who became not only one of my closest family members and friends, but also my guide, teaching me more than I could ever have expected about interspecies communication, intelligence, loyalty, strength of character, and about bonds of love so powerful they span the distance between self and other, even when that divide is as wide as the gulf between species. I have had the joy of sitting in the company of a certain oak tree, a certain stream in a park. Of planting and watching grow the vibrant mulberry tree that feeds me every spring and watches over my tiny, postage-stamp city back yard. These beings and places have taught me, nourished me, loved me. These beings and places are my family and I want to honor and care for them as they have cared for me. I want their health and my own to be ensured for the future. And I want my students to have the chance to experience the sorts of joy, understanding, and growth that come from forging deep, intimate relationships with other beings, and from protecting and loving their world.

Ethicist Anna Peterson (2009) has stated, "we cannot decide whether or not to be connected to other animals and ecosystems. . . . Our choice is not about whether to be connected but about what to do with those connections, how to acknowledge and interpret them" (97). We depend on one another for survival – on other people, other species, and on the ecosystems around us. We must acknowledge these connections in order to survive. But I would like to see us do better. I would like to see us celebrate and explore these connections. And I would like to see us use education as a tool for finding new and better ways to be in the world and to learn about and celebrate our relationships with others.

This is the goal of educating for EcoJustice. As Rebecca Martusewicz, Jeff Edmundson, and John Lupinacci (2011) state, "Rather than being educated to reproduce a culture that we know is doomed to failure, we must begin to educate ourselves and our students about what it means to live differently on the Earth" (7). Following Martusewicz et al. and C. A. Bowers (2001), I believe that education must help students to develop a different way of seeing and living in the world. I believe that it must recognize injustice and exploitation as they are perpetrated against humans, and also as they are perpetrated against nonhumans and the land. It must see these problems as interrelated and as linked to the cultural processes that shape our thinking and therefore our actions. And it must support creative processes of positive reimagining and transformation.

This is what EcoJustice means to me; it is a means to develop a way of living that rejects environmental destruction, cruelty, and oppression as necessary for survival, but instead posits humans as productive contributors to their communities who can nourish and improve their own lives and the lives of others, as well as the land.

To teach for EcoJustice is to demand that students look courageously at the world, think deeply about the forces that have created things as they are, and creatively re-envision what their world can be in the future.

Wendell Berry (1995) has said, "We have tried on a large scale the experiment of preferring ourselves to the exclusion of all other creatures, with results that are manifestly disastrous" (78). I teach because I want to try a new experiment, in thought and in action. I want to see teaching that inspires acts of care and creation, teaching that educates us to desire modes of existence that support the life and integrity of all beings and of the planet. EcoJustice Education has that potential, and I hope this book offers one map as to how we can use it.

The Social Roots of Environmental Problems: Studying Text, Language, and Culture for EcoJustice

How do we make our world into what we want it to be? As members of the delicately interwoven system of living creatures that co-exist on this planet, how do we seek happiness and well-being, not at the expense of other people or other beings, but in ways that support their happiness as well? In every corner of the world today humankind faces specters of looming environmental disaster and social distress, and the sinking knowledge that these specters are of our making. We are each confronted with the question of how to proceed, in our individual and collective actions, in ways that can lead to sustained health and happiness for ourselves and for others in the future.

How have we allowed the devastation to reach this point? How, as David Abram (1997) puts it, have so many of us "become so deaf and so blind to the vital existence of other[s] . . . that we now so casually bring about their destruction" (27–28), and in the process, bring about our own?

In that question we may find the beginning of an answer. Perhaps we are all too often "deaf" and "blind" to the "others" around us, to the other people, cultures, nations, species, lands, and ecosystems with which we share the earth. And perhaps, in attempting to find better ways forward, we must begin, not by debating the specifics of a manufacturing process or a law, but by investigating this deafness and blindness, by examining the motivations and modes of thinking that lead us to act on and in the world, for better or for worse.

Think of the earth. How would you describe it? What imagery, what metaphor, would you use to encapsulate the functioning of the planet, and your own role within it? Do you see the planet as a complex machine and humankind as a mechanic? Do you envision the world as a vast stock of raw materials, waiting for humans to make into objects of value? Do you imagine the earth as a beautiful painting to gaze at and admire? As a war between wild and civilized? As a horn of plenty? As a garden to tend? As a web of relationships? As a neighbor, a community, a home? As a mystery for humans to unlock? As a family of related beings?

Each of these metaphors implies certain assumptions about the purpose of the planet, the purpose of humans, and how we should think about and act toward the larger world.

If we see the world as a collection of objects for our use, as a stockpile of resources, we are likely to view our role as one of extraction or acquisition, and to believe that the key questions we must ask are ones about pace and method of extraction, and about mode of distribution of these resources. We may ask ourselves if the rate at which we are using our stockpile of materials will deplete it too quickly, but we are not likely to question the rights and subjectivity of the beings with which we share the planet, since by this metaphor they are framed as inanimate raw materials or at best as irrelevant bystanders or obstacles.

If we see the world as a garden, we are likely to conclude that we should care for it and tend it, but perhaps also that we stand in a position of authority in relation to it, that it is our right and responsibility to make decisions about its content, that it belongs to us.

If we see the world as our home, we may focus our attention on maintenance and the notion that we should not despoil our own beds. Envisioning the world this way may also raise the notion of heritage and the familial history of birthplace and personal growth. It may encourage a feeling of belonging, and may lead us to think of the earth as our shelter, our protection. However, we may also conclude that the planet is ours to remodel or redecorate at will, and perhaps that there may someday exist an opportunity to relocate to another, grander home.

If we see the world as a family or community, we may think of the welfare of other living creatures, and of the land itself, as linked to our own. We may include other beings in our sphere of concern, and even feel ourselves as part of one another, associating our own identity with theirs. As Wendell Berry describes it, "When we include ourselves as parts or belongings of the world we are trying to preserve, then obviously we can no longer think of the world as 'the environment' – something out there around us. We can see that our relation to the world surpasses mere connection and verges on identity" (1995, 75).

Each of the views described above offers a different perspective for relating to the world, implies a different set of assumptions about ourselves and about other beings. Each one draws our attention in certain directions, highlighting some questions and obscuring others. Each directly and indirectly shapes the ways that we think about the earth, influencing what we value and what behaviors we deem appropriate. Each of these metaphors, and many other belief systems and conceptual frameworks, are at work in our culture at every moment, shaping our thinking and informing our behavior.

In this way, the actions we take in regard to each other and to the larger world are in large part a matter of perception, of belief, of understanding. They are about how we see the world, and how we approach our role within it. Some views of the world direct our attention toward relationality and mutual co-existence, others encourage a focus on autonomy and exploitation, denying shared experience and encouraging "deafness and blindness" to others. Some highlight commonality, others emphasize difference. Some guide us to see the world as full of living, sensing creatures, others cast these beings as objects and commodities. Depending on the framework we adopt, we may perceive the world as alive with interaction and dialogue, or we may hear only silence.

It is for this reason that closely examining our culture has become essential for addressing both environmental and social problems. As Martusewicz et al. (2011) put it, "the ecological crisis is really a cultural crisis – that is, a crisis in the way people have learned to think and thus behave in relation to larger life systems and toward each other. It can be shifted if we learn to think differently about our relationships to each other and to the natural world" (8).

Whether knowingly or unknowingly, we each adopt a combination of the cultural elements available to us to forge our own conceptions of the world, to create our stories, our beliefs, our sense of self and other. And every choice we make is informed by these conceptions we have forged for ourselves. These are the ways we have learned to think and behave; learning to think differently requires recognizing the patterns of thought we have absorbed from our culture and considering how they lead us to behave in relation to others.

Such an undertaking requires the resources of education, including the resources of humanities-based education: critical thought, cultural analysis, historical perspective, understanding of how language and discourse influence us. As a society we must learn to critically evaluate our cultural influences, the thinking, values, ideologies, and narratives that we draw upon to understand ourselves in reference to others and to the world, in order to determine the potential ramifications of various patterns of thought, and to make informed decisions about what we choose to believe. We must learn to recognize our behavior and thinking as constitutive of meaning, and we must choose how to create and re-create ourselves in order to seek out the most positive, nourishing, and sustainable approaches to our world. We need, as ethicist and religious scholar Anna Peterson puts it, to learn to "see the structures in everyday life and then analyze and transform them" (2009, 62).

Ernest Callenbach has stated, "You often gain a new perspective on a value if you see what its concrete consequences are" (2005, 47). Today humankind is coming to see, in dramatic scope, the consequences of the ways of viewing the world that have assumed dominance in our minds and cultures. We see the destruction and abuse that human societies perpetrate daily on other beings, on ourselves, and on the living land. Now we must be willing to see the roots of this destruction in our own minds, to investigate the cultural influences that shape our perceptions of the world. Perhaps, then, it will be possible to identify and adopt better approaches. Perhaps we can learn to recognize and embrace conceptual frameworks that allow us to feel compassion, care, and connection with others, both human and nonhuman, to value interdependence, to see the world not as object but as a living whole, not *for* humans but *with* humans – as David Abram puts it, as "more than human."

Callenbach reflects, "At some great turning points in history, dominant values become exhausted or problematic, and people work out new values that they hope will enable them to survive better" (2005, 47). This must be our mission. We must develop the insight to pinpoint those values that have become problematic, the willingness to contest, reject, or modify them, and the creativity to formulate new values that will enable us to survive better in the future. We must seek a mode of conceiving the world which guides our actions toward justice and sustainability, and which does not exclude others from conversation and consideration.

In order to generate approaches that may lead us toward sustainable attitudes and behavior, we must learn to systematically investigate our own belief systems, evaluate their implications, and work to enact shifts in conceptualization that result in positive interactions with the larger world. And I believe that education is a key site of opportunity to cultivate such methods of critical examination and reinvention. Educators are entrusted and privileged with the opportunity to help students formulate their worldviews, exposing them to the narratives and ideas which allow them to view their embedded conceptions from new angles and to consider new possibilities of thought and action.

This can – and if we want to live in healthier and more just ways, must – be a core purpose of education. Martusewicz et al. say it well: "the purpose of public schools ought to be to help develop citizens who are prepared to support and achieve diverse, democratic and sustainable societies because these are keys to our very survival" (2011, 8).

Further Reading

To read more about the cultural underpinnings of environmental problems and the ideas that inspired and guided the lessons in this book, I highly recommend the following:

EcoJustice Education: Toward Diverse, Democratic, and Sustainable Communities by Rebecca Martusewicz, Jeff Edmundson, and John Lupinacci (2011)

This book is an excellent companion to my own. Martusewicz, Edmundson, and Lupinacci explain in depth the role of belief systems, metaphors, and dynamics of gender, race, and class in influencing issues of justice and sustainability.

The Spell of the Sensuous by David Abram (1997)

A foundational text in considering why human relationships with the natural world have changed over time. Abram builds compelling arguments and suggests that written language played a central part in redirecting human attention away from participation with the natural world.

Everyday Ethics and Social Change: The Education of Desire by Anna Peterson (2009)

A beautiful and refreshing discussion of living life more ethically and lovingly, not through isolated rules but through drawing our ethics from lived experiences of care for those we love.

"Towards Ecopedagogy: Weaving a Broad-Based Pedagogy of Liberation for Animals, Nature, and the Oppressed People of the Earth" by Richard Kahn (in the *Critical Pedagogy Reader*, 2009)

Kahn gives a terrific overview of how environmental education has developed in the U.S. and analyzes what has been missing from environmental education in the past and what education must do better in the future.

Educating for Eco-Justice and Community by C. A. Bowers (2001)

A founder in the field of EcoJustice, Bowers outlines elements he believes are essential in an EcoJustice curriculum. He argues that education for EcoJustice should encourage students to critically interrogate tradition, science, and language.

Composition and Sustainability: Teaching for a Threatened Generation by Derek Owens (2001)

Owens demonstrates how teachers can put composition to work in the pursuit of sustainability, by sharing a writing curriculum designed around issues of place, work, and future.

Teaching for EcoJustice in a Traditional Classroom: How EcoJustice Can Help Fulfill Standard Educational Goals

The goals of EcoJustice education that I discuss in this book are not just goals for an "environmental education" class – they can and should be incorporated across the curriculum. Teaching for EcoJustice means increasing students' critical understanding of their world and of the socio-environmental, cultural, and discursive processes at work around them. But these aren't the only skills that can come from teaching EcoJustice-oriented materials such as the ones in this book. The lessons in this book require students to reflect on cultural beliefs, analyze established assumptions, and consider arguments outside the norm, and in the process they require students to engage in many core educational skills, including to read and write critically, to produce digital media products, to interpret literature, poetry, and art, to study historical events and perspectives, to analyze economic, political, and ethical arguments, and to conduct research.

Because they rely on so many core skills, the lessons in this book can be integrated into traditional academic subject areas including English/Language Arts and Social

Studies at the high school level or a range of humanities-based courses at the college level. I have taught these materials, and have shared them with colleagues who have taught them, in different combinations and at different scales in college courses for departments including American Studies, Media Studies, and English and at levels from introductory to advanced. (For more information on my design and testing of these materials, see Turner, 2011.) Titles of my own courses using these materials have included: Sustainability in American Culture; American Environments: Landscape and Culture; Media Influences on Environmental Discourses and Action; Introduction to Media and Communication Studies; American Food; and Made in America: Material Culture in the United States. Colleagues have taught these materials in developmental reading and developmental writing courses at a community college, adjusting them to be appropriate to the skill level of students who are working to enter introductory-level community college courses. And before teaching these materials in their current form, I began developing the strategies and content while teaching English at a public high school in Baltimore City, working with students in an underserved, low-income neighborhood. In each of these contexts, the materials can be tailored not only to meet the skill-level of the students, but also to meet the requirements of the course or school system, fulfilling mandated objectives and giving students practice with essential content area skills.

To make it as easy as possible for teachers to use these materials in traditional classroom environments with a standard content area focus, I have offered a quick reference at the beginning of each lesson highlighting some key content area skills that are utilized in the lesson – this may be reading and analyzing literature, composing creative writing, interpreting the significance of historical events, conducting research, producing digital media products, etc. In the online resource that accompanies this book, you'll also find lists of current state and national educational standards, including the Common Core Standards, that are fulfilled by each lesson. Without delving into any commentary on Common Core and related standards, I offer them knowing that the standards-based environment of many school systems makes using new materials in the classroom a challenge. Each lesson in this book meets a number of Common Core and other standards, and I hope this will help make it easy for teachers to incorporate them into school systems that require lessons to be linked to these standards.

For the same purpose, it's also worth noting that a number of studies have shown that using socio-environmental issues as an "integrating context" in classrooms can improve student outcomes. The National Environmental Education and Training Foundation has said:

> SEER research since 1997 has shown that environment-based education improves academic performance and learning across the board, regardless of socioeconomic or cultural factors (Hoody, 2002). Indeed, environment-based education appears to be a kind of educational equalizer, improving reading, science achievement, and critical thinking skills across ethnic and racial groups.
>
> *(Coyle 2005, 75)*

These improvements occur not only in science courses, but also in subject areas that deal directly with language, text, and culture. An influential report by the State Education and Environment Roundtable found similar results while studying the benefits of "Using the Environment as an Integrating Context for Learning (EIC)" (Lieberman and Hoody 1998a). In the report, researchers Gerald Lieberman and Linda Hoody state:

> The observed benefits of EIC programs are both broad-ranging and encouraging. They include: better performance on standardized measures of academic achievement in reading, writing, math, science, and social studies; reduced discipline and classroom management problems; increased engagement and enthusiasm for learning; and, greater pride and ownership in accomplishments.
>
> *(1998b, 1)*

The report discusses benefits of EIC within traditional educational disciplines, including language arts. On this subject Lieberman and Hoody state:

> All 17 comparative studies of language arts achievement data found that standardized measures affirm the academic benefits of EIC-based learning for reading, writing, and general language skills. On the average, the EIC students outperformed their peers from traditional programs at all nine of the schools that conducted the analyses.
>
> *(1998b, 4)*

The study found similar results in social studies (7). These studies offer extensive evidence to suggest that educational programs incorporating environmental topics can enhance student learning, comprehension, and skill levels in numerous areas of importance. Educational materials that pursue EcoJustice take this even further, increasing students' capacity for critical thinking and social reflection. This makes EcoJustice education a valuable tool in all school systems and classrooms.

Pedagogy in the EcoJustice Classroom

Teaching for EcoJustice means encouraging certain habits of mind and of heart: the willingness to examine oneself and one's society, to feel empathy for others, to ask difficult ethical questions, and to trace the impacts of language, media, and belief. It means acclimating students to being active participants in discussion and in cultural reformulation. Doing this isn't just about using particular materials, it's about creating a particular atmosphere in the classroom as well. I recommend that, when using the lessons in this book, you pair them with instructional techniques that prioritize dialogue, inquiry, collaboration, and creativity. My materials arise from a critical tradition that posits students and teachers as co-creators of new knowledge and understandings, and I hope that you use them in this spirit.

Think of your classroom as a community of inquiry, one in which your students work together, as they work with you, to explore complex questions. These lessons are intended to create an environment in which students think deeply, looking inward and outward for answers; students should shine a light on their own beliefs and habits, and on the messages they receive from the dominant culture that surrounds them. But this reflection must be grounded in rich analysis of assigned readings and cultural texts. It is important that students' journeys do not end at their own opinions. To achieve this, I try to prioritize both honest discussion and rigorous analysis. I don't expect students to share the opinion of any given author, or to share my opinion, but I do expect them to take the opinions they are exposed to seriously, to give them due thought, and to be willing to question their own long-held assumptions.

It's also important that these lessons offer students space to do emotional work, as well as intellectual work. The subject matter of these lessons deals in bliss and horror, from the deepest connections between beings to the worst of cruelty and oppression, from beauty and awe to devastation. Students should be comfortable *feeling* in response to the materials; this is not only acceptable but also essential. Our society can sometimes send us messages that we shouldn't feel something too deeply – that it is weak, or foolish, or childish to love nonhuman animals, to grieve for a forest when it is cut down, to be angered by the treatment of animals in factory farms. It's important to acknowledge that these norms can serve to keep us isolated from other beings and to prevent people from objecting more strongly to the destruction and pain they see around them. Feeling for and with others, recognizing their pain as our own, feeling harm done to the land as harm done to ourselves – these are not abnormal responses, they are simply ones that many of us have come to repress or forget. A classroom is a place where students wrestle with the realities of the world, both bleak and beautiful; it should be a place where it is safe and encouraged to feel deeply and to discuss those feelings. And when teaching for EcoJustice, relearning to feel empathy, compassion, interdependence, and connection can be both a revolutionary act and a way of liberating students, giving them the freedom to be honest with themselves and others about their experiences and their desires for a better world.

Structure of the Lessons

Each lesson in this book starts with a short description of the lesson, a list of the readings and other materials I assign as part of the lesson, and a narrative explaining my procedure for teaching the lesson. I also include the text of writing assignments I assign, and examples of student writing completed in response to these assignments. The procedure I describe for each lesson is not meant as a script, but as a model that will give you a clear picture of how I engage with students around the ideas and skills of the lesson. I've titled it "My Procedure" to remind you that I am sharing my own experience, and to leave you space to create your own experience with the materials. The lessons in this book are designed to be flexible. They can be adjusted to include fewer or more readings, and the procedures can be modified to suit the needs of particular students and classroom settings. But they are also designed to prioritize critical thinking and active student participation, and however they are modified, I encourage teachers to maintain these priorities. The goal of the lessons is to get students thinking deeply about their world and developing capacities that help them act for positive change. There is no one correct opinion, as long as students are grappling with questions raised by these lessons, reflecting on their own beliefs, and making an effort to step outside normative assumptions and consider alternative viewpoints and alternative future paths.

The lessons in this book are divided into eight units. You can teach as many or as few of these units, and the lessons within them, as you like. Some lessons build explicitly from previous lessons and should be taught together; when this is the case I've identified that fact in the lesson. If you do teach multiple lessons or multiple units, I recommend teaching them in the order I list them when possible, since they're designed to work together in this order and help students develop cumulative skills and understandings.

Most lessons in this book begin with a set of assigned readings that students complete outside of class, followed by a writing assignment I call a "response paper." This written assignment allows students to process and analyze the readings, preparing them for in-class discussions and activities. Depending on the skill level and needs of your students, you may choose to add more pre-reading activities to the lessons than I use, in order to help prepare students for their readings.

I structure my "response paper" assignment through an online discussion board; after each set of assigned readings and before a given class period, students post their written response papers to an online forum. I then have students post a set number of replies to their classmates' response papers. This gives students the opportunity to read some of their classmates' analyses, often exposing them to different interpretations than their own. The replies also provide a way to extend discussion of the readings beyond class time, and give students who might not have spoken up much in class an opportunity to share their thoughts with each other in a less high-pressure setting.

The response paper assignment is very well suited to this online discussion board structure. However, if your students don't have regular internet access or it's otherwise not feasible to do this as an online assignment, a "double-entry" or "reader response" journal would serve a similar purpose, as would simply having students turn in a typed response paper directly to you.

Below is the text of the response paper assignment I give to students. I include it here, rather than in the lessons that follow, because I use it in every lesson.

Writing Assignment: Response Papers

Each week you must post a response to that week's readings on our online Discussion Board. Responses should include your personal reflections as well as critical analysis of the readings. You must directly discuss at least two of the readings for that week in your response. Responses should be approximately 500–700 words. If you quote directly from the readings (which is encouraged), make sure you cite page numbers.

Some questions you may consider when you write your weekly response papers:

- What is your favorite passage, sentence, or line from the readings and why?
- What struck you most about the readings? What surprised you? What new insights did you gain?
- Do you agree or disagree with the authors' points? Why?
- Are there any passages or concepts in the readings that confused you? Points you want to clarify? Questions you want to discuss? Complex issues you want to unpack?
- How do these readings relate to other readings we've done in class? Do these authors make points that expand, reinforce, or contradict other pieces we've read?
- How can you relate these readings to your own life experiences and to texts you've read and seen outside of class?
- How do the readings relate to this week's topic? Do these readings give you any new perspectives on the topic we're discussing this week, and/or on sustainability in general?

Along with this assignment I include a requirement that students reply to their classmates' response papers.

Writing Assignment: Response Paper Replies

Each week you must reply to at least two of your classmates' posted response papers. Replies may include asking a question, expanding on a classmate's idea, agreeing or (politely!) disagreeing with a classmate's point (and explaining why), relating a classmate's post to your own post, etc. Replies should average at least 100 words.

In a college setting, with classes that meet twice a week, I typically schedule these papers and replies as such: I assign readings once a week, to be completed before the first class session of that week. Before that first class session, by a set time, students must post a response paper in which they discuss that week's readings. Prior to the second class period of the week, students must reply to two of their classmates' papers. In settings where classes meet every day or more frequently, like some high schools, you could adjust this pacing as needed. You may, for example, choose to engage in pre-reading activities and introduce assigned readings on a Monday, have response papers due Tuesday, discuss the readings in class on Tuesday and Wednesday, have replies to classmates due Thursday or Friday, and then introduce any additional activities for the lesson that students could work on in class Thursday and Friday and/or over the weekend.

Most of the readings I assign in these lessons will be accessible to students with moderate to strong reading skills, and some are accessible to students with limited reading skills. At the end of the book I list all the readings from all of the lessons, and in this list I include a short passage from each reading. This should help give you a sense of the content and difficulty level of the readings. In the online resource that accompanies this book, I include the text or links to the text of some of the readings, as well as some additional links to other articles, videos, and content that could be included in the lessons. You will occasionally find a reading in one lesson that is duplicated in another lesson; if you teach all of the lessons in the book in order, you may assign the reading the first time it appears and leave it out of the second lesson in which it appears if you wish. In these cases I listed the reading in more than one lesson so that if you choose to teach one lesson and not the other, the readings essential to each lesson will be included.

As I teach these materials myself, I sometimes use additional assignments that are not listed in the lessons because they are ongoing, not unique to a specific lesson. For example, I may assign students to lead discussion on a certain set of readings, developing discussion questions and facilitating the class session. Or I might have students do background research on the historical context of a set of readings, or read an outside book that's related to the topics we're discussing. I consider these general-practice assignments, and most teachers will certainly have similar practices they already employ – please feel free to add such activities into any of the lessons in this book. For more detail on these assignments and other activities I sometimes add to my lessons, see the online resource for this book: www.routledge.com/9781138832923.

I also strongly encourage teachers to link these lessons to their local regions and communities when possible. I've given a few examples of how I do that in some of the lessons, and more ideas can be found in the online resource.

At the end of each unit is a section titled "Effect of the Lessons." In these short sections I reflect on key skills and qualities that I believe are cultivated by the lessons in that unit (and by the curriculum in general). I consider these skills essential to the practice of EcoJustice, and I encourage you to keep them in mind as you teach the lessons and to think about ways that you can help your students to develop them.

As I describe my process for teaching these lessons, you'll notice that I share a number of quotes from my students. These serve as samples of good student work that I've received in response to my assignments, quoted verbatim from students, without alteration (except to cut passages for length). The student work that I quote from in this book is among the more exemplary work submitted by my students over the years, but for the most part it is not rare or unusual. I've received many comments from students that are extremely similar to the ones quoted here – in many cases I had so many wonderful pieces of student writing to choose from that I was distraught over which passages to share! And even when students don't produce written work of the same quality quoted here, they often express many of the same sentiments. Some of the pieces of student writing I've selected to use in the book I chose because they were very special, offering a striking insight or a beautiful piece of writing, but many of them are representative of common comments made by my students in response to these lessons and assignments.

These quotes from my students are meant to provide examples of the sorts of insights and analysis to look for from students as you teach these lessons. But I hope that they do something more, beyond this; it's my hope that including the voices of my students helps to manifest my pedagogical stance more clearly than simply describing it. These lessons should be about dialogue, about sharing questions and insights and revelations together. My students have at times raised points about the texts I've assigned that I had never realized myself, they've asked profound questions and pointed me and the class as a whole toward new ideas. While I designed these lessons, they are really a shared product, and one that my students played an important role in. I hope the experience of reading this book will hint at what the experience of teaching these materials should be – a blend of ideas shared by authors, poets, filmmakers, and artists, ideas shared by the instructor, and ideas shared by the students themselves, coming together in the spirit of honest inquiry and creative rethinking.

A Note about Terms

When trying to speak in respectful and positive ways about other species, ecosystems, and the biosphere, figuring out the best terms to use can be tricky. You'll notice that I use several different terms in this book. To some extent I've tried to mirror the language I use with my students, to provide a sense of how I navigate my terminology in the classroom. I find that, as my students grapple with the question of how to modify their language use to achieve the most sustainable outcomes, they can get rather overwhelmed at the range of new terms suggested to them in their readings. So when modeling language choices I try to mix some less familiar with some more familiar and relatable terms. I often use the phrase "natural world" to describe the entirety of the living planet beyond the human-built environment. The word "natural" is problematic and slippery at best, but this term is simple for students, and I find it preferable to "nature," a rather monolithic and difficult to define word. Some might point out that humans, too, are "natural" and some people feel that using the phrase "natural world" to describe the world beyond that which humans have built is therefore inaccurate. But for the purposes of working with students, I think this distinction is a red herring. My students are quite capable of recognizing that humans are part of the "natural world" and still using that term in ways that are clear to me and to their classmates.

A term I like very much for describing the larger world beyond that which is human-centered is David Abram's phrase "more-than-human world." You'll notice that I use this phrase at times throughout the book, however, I limit my use of it with my students since I find it to be somewhat less accessible for them as they start their journey of relanguaging.

Another complex decision is how to speak of other species of animals besides humans. In English we usually simply call other species "animals." However (as the readings in Unit 2 point out), this creates a false dichotomy between humans and other animals, since humans are also a species of animal. I most frequently use the term "nonhuman animal" to discuss other species. This term is also a compromise, since it still refers to humans as the standard that other animals are being compared to. But I consider it to be the most functional option, and my use of this term in my classroom still expands students' awareness of our typical habits when referring to other animals and reminds them of the false distinctions that are often drawn between humans and other species. I find by the end of the semester many students have chosen to adopt the term themselves.

I encourage you to be thoughtful about the language you use with your students, discuss your choices with them, and be prepared to have an open conversation about the positive and negative implications of each option. You may not land on any terminology that you feel completely content with, but you'll be much more aware of the motivations and consequences of your language use, and that is an enormously valuable outcome in itself.

Getting Started

Now it's time to begin the lessons. I'm honored for these materials to be used in the classrooms of teachers and future teachers. I hope these lessons will offer you and your students experiences of insight and inspiration, and that they can play some role in helping each of us make the world what we want it to be.

References

Abram, David. 1997. *The Spell of the Sensuous: Perception and Language in a More-Than-Human World*. 1st Vintage Books Ed. New York: Vintage.

Berry, Wendell. 1995. *Another Turn of the Crank: Essays*. Washington, DC: Counterpoint.

Bowers, C. A. 2001. *Educating for Eco-Justice and Community*. Athens: University of Georgia Press.

Callenbach, Ernest. 2005. "Values." In *Ecological Literacy: Educating Our Children for a Sustainable World*, edited by Michael K. Stone and Zenobia Barlow, 1st ed., 45–48. San Francisco: Sierra Club Books.

Coyle, Kevin. 2005. *Environmental Literacy in America: What Ten Years of NEETF/Roper Research and Related Studies Say about Environmental Literacy in the U.S.* Washington, DC: The National Environmental Education & Training Foundation.

Kahn, Richard. 2009. "Towards Ecopedagogy: Weaving a Broad-Based Pedagogy of Liberation for Animals, Nature, and the Oppressed People of the Earth." In *The Critical Pedagogy Reader*, edited by Antonia Darder, Marta P. Baltodano, and Rodolfo Torres, 2nd ed., 522–540. New York: Routledge.

Lieberman, Gerald A., and Linda L. Hoody. 1998a. *Closing the Achievement Gap: Using the Environment as an Integrating Context for Learning. Results of a Nationwide Study.* San Diego: State Education and Environmental Roundtable.

———. 1998b. *Executive Summary: Closing the Achievement Gap: Using the Environment as an Integrating Context for Learning. Results of a Nationwide Study.* San Diego: State Education and Environmental Roundtable.

Martusewicz, Rebecca A., Jeff Edmundson, and John Lupinacci. 2011. *EcoJustice Education: Toward Diverse, Democratic, and Sustainable Communities.* 1st ed. New York: Routledge.

Owens, Derek. 2001. *Composition and Sustainability: Teaching for a Threatened Generation.* Urbana, IL: National Council of Teachers of English.

Peterson, Anna Lisa. 2009. *Everyday Ethics and Social Change: The Education of Desire.* New York: Columbia University Press.

Turner, Rita. 2011. "Critical Ecoliteracy: An Interdisciplinary Secondary and Postsecondary Humanities Curriculum to Cultivate Environmental Consciousness." Baltimore, MD: University of Maryland, Baltimore County.

Unit 1

NATURE AND THE SELF

[handwritten: aim: cultivate care for other beings · peace, love, compassion, empathy]

Free thinker! Do you think you are the only thinker
on this earth in which life blazes inside all things?
Look carefully in an animal at a spirit alive;
every flower is a soul opening out into nature . . .
"Everything is intelligent!" And everything moves you.
—*Gérard de Nerval, "Golden Lines" (38)*
Lesson 1.1

The lessons in this unit ask students to think about their relationship with the natural world and with the nonhuman beings that co-inhabit this world with them, and to consider both what impact the natural world has on them and what impact they have on the natural world and on other beings. Through poetry, short stories, essays, and visual art, students are presented with examples of important encounters between humans and the natural world, and with details of both positive and negative consequences of human interaction.

Some key questions that the lessons in this unit should raise for students:

- What part does the natural world play in your daily life?
- Are plants, animals, and the land important to you? Why/why not? In what way?
- What relationship *should* human beings have with the natural world?
- Have you ever had an experience in the natural world or interacting with an animal that influenced your life?
- Does spending time in the natural world have benefits for people? What sort?
- What positive or negative impacts do human beings have on other animals, plants, and the land?
- How do other living beings experience their relationship with humans?

[handwritten in left margin: Rubric 1]

LESSON 1.1: LITERATURE AND CONNECTION

Reading, Art, and Reflecting on Bonds with the Natural World

About this Lesson

In this lesson, students read a selection of poems, short stories, and essays in which authors explore experiences interacting with the natural world and with nonhuman creatures, as well as poems in which authors write from the perspective of other beings, imagining what those beings feel and commenting on how those beings are treated by humans. Students also view art that explores human relationships to the nonhuman world. Students think about how these pieces are written and created and what they convey, what they say about the value of the natural world and of other beings, and what they express about relationships between humans and others. Students also write short reflective essays responding to the assigned readings and viewings.

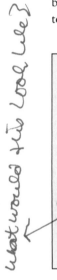

what would this look like?

Lesson Objectives

Students will:

- Analyze and respond to the features of literary prose and poetry.
- Reflect on the role of literature, poetry, and art in exploring relationships between humans and other beings, and between humans and the land.
- Reflect on positive relationships between humans and other beings, and between humans and the land.
- Employ empathy and imagination to understand personal experiences of connection, affection, and personal growth through interaction with the nonhuman world.
- Critically analyze literature, poetry, and art.

Lesson Activities at a Glance

1. Assign literature and poems for students to read, split into two sets of readings.
2. Students write response papers outside of class.
3. Conduct in-class discussion in groups and as a whole class, discussing the content, message, and features of each passage and the experiences with the natural world described in the readings.
4. View selections from documentary film and discuss student reactions.

Key Content Area Skills: reading and analyzing literature and poetry, writing analytical and reflective essays.

Texts Used in this Lesson

Poems

Gérard de Nerval, "Golden Lines," in *News of the Universe: Poems of Twofold Consciousness*, 38

Leroy V. Quintana, "Sharks," in *Poetry Like Bread*, 202–203

Ferruccio Brugnaro, "Don't Tell Me Not to Bother You," in *Poetry Like Bread*, 75

Genny Lim, "Animal Liberation," in *From Totems to Hip-Hop: A Multicultural Anthology of Poetry Across the Americas, 1900–2002*, 34–36

Federico García Lorca, "New York (Office and Attack)," in *News of the Universe: Poems of Twofold Consciousness*, 110–112

Jimmy Santiago Baca, "Ah Rain!" in *Poetry Like Bread*, 54

Wisława Szymborska, "The Silence of Plants," in *Poems: New and Collected 1957–1997*, 269–270

Wisława Szymborska, "Among the Multitudes," in *Poems: New and Collected 1957–1997*, 267–268

Rainer Maria Rilke, "Ah Not to Be Cut Off," in *Ahead of All Parting: The Selected Poetry and Prose of Rainer Maria Rilke*, 191

Rainer Maria Rilke, "I Live My Life," in *News of the Universe: Poems of Twofold Consciousness*, 76

Tupac Shakur, "The Rose That Grew from Concrete," in *The Rose That Grew from Concrete*, 3

Essays and Short Stories

Passages in *Reading the Environment* edited by Melissa Walker:

"Walking" by Henry David Thoreau, 41–44
"The Serpents of Paradise" by Edward Abbey, 51–57
"A Blizzard Under Blue Sky" by Pam Houston, 57–62
"Living Like Weasels" by Annie Dillard, 63–66
"The Call of the Wild" by Gary Snyder, 71–73

N. Scott Momaday, "The Man Made of Words," in *Our Land, Ourselves*, 71–73

Aldo Leopold, *A Sand County Almanac*:

"January Thaw," 3–5
"The Green Pasture," 54–56
"If I Were the Wind," 70–71

Terry Tempest Williams, *An Unspoken Hunger: Stories from the Field*:

"The Architecture of a Soul," 13–15
"Redemption," 143–144

Visual Art

View works of Andy Goldsworthy (see online resource for links).

Documentary Film

Rivers and Tides: Andy Goldsworthy Working with Time

Note: Publication information for these materials is listed at the end of the unit and in the full List of Readings at the end of the book. To help you get a better sense of the readings, I have also included short passages from each one in the List of Readings at the end of the book. The text of some of the readings, or links to find the text, is available in the online resource.

My Procedure

I typically begin this lesson by asking students to share any worries or questions they have about the environment or about humankind's relationship with the natural world. Since this lesson is often the first in the semester, starting with this question helps introduce my students and gives me a sense of what's on their minds. I then share a slideshow in which I raise some questions of my own about humankind's relationship with the natural world. This slideshow is available in the online resource. Alternatively, you can use the key questions listed in my introduction to this unit as inspiration for a pre-reading activity.

After talking about our concerns over environmental problems and how humans interact with the natural world, I assign the readings listed above. In these readings authors describe meaningful experiences they've had interacting with the natural world, they reflect on harm done to animals and the planet, they consider the possibilities for communication with other species, and in some cases they explore the perspective of other beings, writing from the point of view of an animal or of the earth. I typically conduct this lesson over two weeks, splitting the readings: first I assign the poems, having students write about them and discuss them in class for one week; the following week I assign the short stories and essays, as well as links to view the visual art. In class during the second week I also show parts of the documentary *Rivers and Tides*. This lesson could easily be conducted over a longer stretch of time than two weeks, or fewer or more pieces could be assigned to modify the time span.

Students read the assigned texts outside of class, and then write a response paper reflecting on the texts, which they post to an online discussion board as part of an ongoing assignment for the class. Each student posts a response paper online weekly, and they write replies to each other's posts. This allows students to think through and share their initial reactions to the materials prior to discussing them in class, and it encourages an atmosphere of open conversation and inquiry, as students write back to one another's thoughts. (For more details and for the text of this assignment, see the Introduction.) For this lesson, since I break the readings into two groups over two weeks, my students write two response papers – first they write about the poems, and then about the short stories, essays, and visual art.

Below you'll find a good example of a full response paper written for this lesson, by a student of mine who wishes to remain anonymous.

Student Writing: Response Papers

From the choice of poems this week, "Golden Lines" by Gerard de Nerval and "Sharks" by Leroy V. Quintana were the two most enjoyable in my opinion. "Golden Lines" immediately captivates the reader with the intense line, "Free thinker!" with a strong voice as the poem begins a strict lecture to the audience, the human species. To me, the poem calls out the audacity of humans to perceive themselves as superiors to everything. The author uses a quote from Pythagoras in line 8, "Everything is intelligent!" as a reminder that life takes place in many forms and different forms play different yet significant roles.

The word choice the author employs emphasizes the intensity of life and spirit that if searched for carefully can be found in any "substance of creation" (line 10). For example line 2 says, "Life blazes inside all things". The word "blazes" makes me picture a burning, passionate soul lit inside the body that radiates a positive energy. I agree with the poem that animals have a soul and sprits as well as plants. However, I have yet to fully agree/grasp the notion that materials such as metal and minerals contain some "touching love" or other spirit (line 7). On the other hand, I do concede that the land, soil, and earth-produced materials play an important role in the circle of life; they are habitats, enrichment, and structure that support other humans, plants, and animals. This overall connection between life is nicely stated in the poem; "'Everything is intelligent!' And everything moves you" (line 8).

Another aspect of the poem that made me think was the use of the words "creature" and "Holy Thing". In my opinion, they allude to the creation of life from God's hands. This leads to the question: is there a distinction between a creature and human? Or are they the same? Usually, "creature" holds a negative connotation, linked with a mysterious or savage animal. In contrast, "Holy Thing" seems pure and good. The author notes in line 12 that "Often a Holy Thing is living hidden in a dark creature" which leads me to conclude that there is more than what meets the eye. We should not simply overlook what is visible because the physical shape may be misleading and the spirit of life is not easily visible to the arrogant "free thinkers" who hold no consideration for anything else. Is it not ironic that the free thinkers are so limited in their thoughts to believe they are the only thinkers on this earth (lines 1–2)?

Moving on to a different perspective, "Sharks" is narrated in the point of view of a shark. Again in this poem, the readers see the negative consequences that follow limited compassion and the belief that humans are the only

(continued)

(continued)

free thinkers. Sharks are stereotypically portrayed as a fierce, intimidating, and hostile predators that roam the deep oceans. The poem leaves the message that sharks are only seen as a market good or trophy prize to hand on the walls to boost egos and decorate. Their skin, fins, and jaws are used to make profit without second thought to their life. And once they are caught, killed, and sold, their waters are continuously polluted. One line that stands out is from line 5. The goods that sharks produce are "all vital for the survival of their countries." This demonstrates the greed of consumers who care more about profit and the state of the market rather than the shark itself, stripping away any sense that the shark is a living being. Overall, these two poems enlighten the reader to look in multiple perspectives and not just for the benefit of humans/self.

In this response paper, my student analyzes two of the poems from this lesson that encourage readers to think about perspectives beyond their own, and to consider the value of other beings. These are key behaviors that the poems in this lesson should encourage. Many students respond strongly to "Sharks," one of the poems my student writes about in this example. "Sharks" is narrated from the perspective of the shark. The piece reflects on the way that humans misrepresent and exploit sharks, and compares this sort of exploitation to that experienced by undocumented immigrants. Here's an excerpt from another response paper which provides another good example of a student discussing this poem, and beginning to question the assumptions embedded in our dominant human perspective:

Student Writing: Response Papers

by Rebecca Postowski

It was interesting to view environmental issues from the creature's point of view and how it compares to a human being's perspective. A line from the poem that really spoke to me was, "When men purchase suits from our skin they become dukes, barons of giant corporations" (Quintana, pg. 202 lines 1–2). The shark's death is being glorified to humans, as the carcass is used to promote a higher social ranking and status. It's interesting to me to notice how various attitudes from individuals effect interactions with the environment. While sharks tend to be portrayed in the media as "vicious" or "deadly," aren't humans being the "deadly" ones for killing innocent sharks?

Other poems also challenge students' assumptions about our relationships with other species. In "The Silence of Plants," Wisława Szymborska questions how we can relate to beings that are very different from ourselves. Here is an example of a student wrestling with this question in reaction to that poem:

Student Writing: Response Papers

by Jacob Rosenborough

The piece displays the fact that although we have a lot in common with plants we don't communicate with them. At the surface, I thought this was quite simply because plants don't talk *duh! right?*. But as I thought about it more I realized it might be backwards. Maybe it's that humans don't listen? I came to this idea because, as in "Don't Tell Me Not to Bother You", humans don't want to be "bothered". But plants have a lot to offer us humans, as Szymborska says: "we're traveling together" and "talking with [plants] is essential". Plants feed us (and they feed other things we feed on), they can teach us significant amounts of knowledge involving biology and chemistry, and they can be useful in medicine. The bottom line to this idea of mine is: Environmental science shows that every part of the environment works together, including plants and humans, so maybe plants aren't "silent" towards us, but we just think that they are because of how disconnected to the World we all are.

After students have read this first set of selections and they have posted their online response papers, it's time to discuss the materials in class.

When we arrive in class, I break students into groups of two to four people, and I assign each group a poem to analyze in more depth. (I typically split people up based on which poem they wish to discuss in their groups, or let groups weigh in on the selection.) I ask each group to discuss the details of the poem and any questions they have about it, to determine the basic events that take place in their poem, and to attempt to identify the narrator. I then ask them to answer the following questions, identifying particular lines, words, and phrases in the poem that support their answers. I move between groups helping to guide and support their discussion.

Questions for discussing the poems:

- What specific imagery and language does the poet use to describe nonhumans and the natural world? What is its effect?
- How would you describe the poet's feelings toward the natural world, and what leads you to that conclusion?
- If the poet is writing from the perspective of another being, what details does she convey about life as this being? How can you tell who this being is and what she/he/they feels?
- If you could select one line that best expresses this poem?
- Why do you think the poet created this text? What was her goal or purpose? What work does the poem do?
- Do you identify with the poet or subject of the poem in any way, or did the text change anything about your attitudes or feelings?
- What value (if any) do you think this text has for our culture? Why?

Once they've discussed their assigned poems in groups, I have each group of students share their conclusions with the class. As a whole class, we discuss the poems, exploring particular passages, sentiments expressed, and students' reactions. We talk about which poems each student liked most, any sections they found confusing, their feelings toward the narrators of the poems, any passages they relate to strongly or similar experiences they've had, and whether they can identify with the nonhuman beings whose perspectives are imagined in the poems.

I then assign the second set of readings; students read these outside of class and write response papers. This second set of readings includes short stories and essays in which authors reflect directly and indirectly on relationships they themselves have had with the natural world, and on the relationship modern society has come to have with the natural world. Included in this set of readings are some pieces that offer examples of great benefits humans get from interaction with the natural world, including stories of a woman who was struggling with depression and found insight through winter camping, a woman who bonded with her grandmother while collecting seashells, and lessons learned by a man who formed a friendship with a snake. One passage from *A Sand County Almanac* by Aldo Leopold describes the beauty of a riverbank. Here's an except from a student's response paper in which she discusses the insights she got from this piece:

Student Writing: Response Papers

by Jennie Williams

The piece, "The Green Pasture" presents nature's art as one that can only be viewed once because of the natural shifts and changes in the environment at every moment. The visual description grasped my attention when I read this piece. To imagine the river and the passing wildlife as the artists and the sandbank as the canvas was eye-opening. . . . When it is viewed as an art form instead of simply an environmental element, to protect it is to appreciate the innovative and creative influence of the Earth. . . . "The artist has now laid his colors, and sprayed them with dew" (Leopold, 55). Nature is so very influential to humans in that it provokes creative thought as well as reminiscent inspiration. These pieces remind us that our environment is not only a place for renewable resources and habitat, but it also provides an emotional and artistic outlet where our minds may thrive.

After students have written about this second set of readings, in class I conduct small group or full class discussions about the texts, considering students' favorite readings, examining particular passages, and sharing reactions.

To help facilitate this discussion, I often draw from students' response papers a list of questions they asked about humankind's relationship with the natural world (rephrasing their questions to make them more general). I title this "Questions

You Raised," and usually share it via PowerPoint in class. Here are some examples of questions I have drawn from students' response papers; I put a slide with this exact text onscreen for students:

Questions You Raised

- What rights do we have to interact with and use nonhuman nature?
- What do nonhuman creatures mean to us, and what do we mean to them?
- In what ways do our ideals differ from our everyday behavior?
- What possibilities exist for understanding, communicating, and empathizing with nonhuman creatures?
- Is it important to attempt to communicate with or understand nonhumans? Why?
- In what ways can and should we identify with and think about the nonhuman world in our daily lives?
- In what ways can interacting with nonhumans teach us about ourselves or help us deal with our feelings?
- What insights do these poems/artworks have to offer readers/viewers? What do they have to offer society?

If I feel there are any important questions that aren't represented by the list I've gathered from student writing, I may follow this list with some of my own questions. (See the key questions at the beginning of the unit for examples of questions I consider important – and notice that students have often come to most of these questions on their own, as seen from the above list.)

During this week I also show excerpts from the documentary film *Rivers and Tides,* which follows artist Andy Goldsworthy as he creates some of his nature-based "landscape art." We typically watch at least the first 17 minutes of the film in class, just to give students a chance to hear Goldsworthy talk about his process and watch him work; the whole film is well worth watching, however. Seeing Goldsworthy talk about his work and respond to the natural spaces in which he creates art offers another interesting opportunity for students to consider different types of human relationships with the natural world. Goldsworthy uses only materials he finds in a given place, including driftwood and icicles – which he breaks with his teeth and reattaches to one another to form shapes. In the film he discusses natural features and rhythms that he tries to emphasize in his work, and he speaks of getting to know a place, learning about it, and being inspired by what he finds in each unique location. After viewing the film we discuss it, as well as the pieces of Goldsworthy's artwork students viewed online. Students are often split, some wondering if Goldsworthy should be manipulating natural materials at all and if he has a right to alter an undeveloped place, even if his work is temporary. Others are inspired by his work, by the close observation and affection he demonstrates for the places he works in, and by the potential light his art can shed for those viewing it, possibly helping them to admire the land and feel close knowledge and affection for it as well.

Having discussed these readings and viewings in class, I segue into two creative writing assignments – you will find these in Lessons 2.2 and 2.3. I assign these one after the other, but often overlapping with the activities of Lesson 2.1.

Products and Assessment

Students demonstrate their engagement with the assigned materials in this lesson in two primary ways. Having read the assigned texts, students write response papers analyzing and reacting to the readings. These response papers should demonstrate careful reading and thought about the texts. After completing these papers, students come to class and engage in class discussion. During this discussion they work in groups and as a whole class to analyze poems and prose passages, justify their analysis using specific evidence from the text, and share this analysis with the class. Here again they should be demonstrating close knowledge of the texts and thoughtful analysis. These two activities can be used to evaluate students' achievement of the lesson objectives.

I use the following grading criteria to evaluate student work for this lesson. I have, in different classes, formatted these criteria as a grading rubric or listed them in my course syllabus. How to format and apply the grading criteria will depend on individual context, school requirements, and teacher preference.

Grading Criteria

- Student demonstrates knowledge and critical analysis of the features and techniques of poetry and literary prose.
- Student explores the personal and cultural significance of relationships between humans and the natural world.
- Student thoughtfully reflects on personal memories about experiences with the natural world.
- Student thoughtfully imagines the lives of others, including nonhuman creatures.
- Student demonstrates empathy in imagining the experiences, feelings, and desires of others.
- Student demonstrates thorough reading and comprehension of assigned course texts.
- Student demonstrates personal reflection, critical thought, and insight into course texts.
- Student's arguments are clear, well-developed, and documented with evidence from texts.

- Student's work demonstrates critical analysis of course topics, questions, and subject-matter.
- Style, usage, format, grammar, imagery, and presentation support meaning and are intentional, creative, and original.
- Work meets all requirements of the assignment and is utilized to facilitate development of personal understanding.

*literacy assumption .

* textual representation
used to for assessing.

what about a # spoken word;
oral assessment

~ 8th

LESSON 1.2: WRITING THE CONNECTIONS THAT SHAPE US

About this Lesson

In this lesson, students review relationships with the natural world described in literary passages they read in Lesson 1.1. They consider the sorts of relationships discussed and the benefits expressed by authors. Then students write their own poems and short stories describing a formative event or relationship in their lives that involved interaction with the natural world. In the process, students reflect on ways that their own identities have been shaped by other beings and by the land, and the insights or benefits they have gained from these experiences.

Lesson Objectives

Students will:

- Reflect on the role of literature in exploring relationships between humans and other beings, and between humans and the land.
- Reflect on positive relationships between humans and other beings, and between humans and the land.
- Reflect on their own formative experiences interacting with the nonhuman world.
- Employ empathy and imagination to understand personal experiences of connection, affection, and personal growth through interaction with the nonhuman world.
- Compose original creative writing.
- Use language creatively to reflect on personal experiences and emotions.

Lesson Activities at a Glance

1. Review literary passages assigned in Lesson 1.1.
2. Conduct in-class discussion in groups and as a whole class.
3. Students complete a creative writing assignment.

Key Content Area Skills: writing poetry and creative prose.

Texts Used in this Lesson

There are no new assigned readings for this lesson; review texts from Lesson 1.1.

My Procedure

This lesson builds on the literary passages and poems students read in Lesson 1.1. After reading these passages, writing response papers reacting to them, and discussing them in class, I ask students to reflect further on any similar experiences they have had to those described by the authors. A number of the assigned readings describe moments spent by the authors in which they interact with a nonhuman animal or natural space: standing in the rain, winter camping, collecting seashells with a beloved grandmother, releasing a captive duck, pondering how to relate to plants, watching seasons change. In each case, the authors come away with new insights or changed emotions. In class, I ask students if they have had any similar experiences, and if they could relate to the emotions the authors describe. Many of my students collected shells or rocks as children, explored wooded areas near their childhood homes, or had a memorable moment of meeting or interacting with a wild animal, often feeling a brief affinity and connection.

I ask students to share a bit about some of these experiences in a full class discussion. Such sharing could also be done in smaller groups. In the response papers they previously wrote (see Lesson 1.1) discussing these readings, students may have already shared some of their own memories and experiences – this is a great starting point, as they can recap what they described in their response papers and then elaborate, and other students who hadn't thought to write about their own memories can then join in.

Once this discussion has taken place, I present students with this creative writing assignment:

Writing Assignment

Write a poem or short story from your own perspective, describing an experience you've had with the natural world or a meaningful interaction with a nonhuman being. Describe where you were, what was going on in your life, how this event came about, and what impact it had on you. Think about how you felt while it was going on, and also how you feel about the event looking back on it, and any lasting feelings or insights it has raised for you.

Poems and short stories should demonstrate personal reflection and insight as well as detailed, empathetic consideration of the experiences of others, and should employ creative details, imagery, word choice, rhythm, style, and syntax to engage readers and evoke an emotional response.

I haven't shared an example of my students' writing here, as their writing for this assignment includes personal details about their families, hometowns, and childhood memories. Student writing in response to this piece should be thoughtful and personal. The events students describe don't have to be dramatic, but they should demonstrate genuine thinking about how experiences with the natural world can affect a person emotionally, intellectually, and psychologically. This should be an opportunity to consider how one's emotional growth, mental well-being, and interpersonal understanding may be intimately linked to close relationships and interactions with other beings and with the land. Many students will begin this lesson feeling almost entirely independent of the natural world, but in the course of writing this assignment some come to discover that they have been shaped in ways they didn't realize.

After students do this creative writing assignment, I sometimes ask if any are comfortable sharing what they wrote in class – this is entirely voluntary. We may also discuss what insights students gained from writing these pieces, and if they realized anything new about their own relationship with the natural world.

Products and Assessment

In this lesson students engage in class discussion in which they should be making specific reference to details from course readings and relating those details to their own lives. Through this process they should demonstrate careful reading, comprehension, and reflection. After this, students complete a creative writing assignment. This assignment should demonstrate genuine personal reflection and creativity. Their participation in class discussion and their creative writing can be used to evaluate their achievement of lesson objectives.

I use the following grading criteria to evaluate student work for this lesson. I have, in different classes, formatted these criteria as a grading rubric or listed them in my course syllabus. How to format and apply the grading criteria will depend on individual context, school requirements, and teacher preference.

Grading Criteria

- Student thoughtfully explores personal memories about experiences with the natural world.
- Student uses creative writing to reflect on personal experiences.
- Student employs empathy and imagination to investigate how experiences with the natural world have influenced her/his life.
- Student demonstrates thorough reading and comprehension of assigned course texts.
- Student demonstrates personal reflection, critical thought, and insight into course texts.
- Student's arguments are clear, well-developed, and documented with evidence from texts.
- Student's work demonstrates critical analysis of course topics, questions, and subject-matter.
- Style, usage, format, grammar, imagery, and presentation support meaning and are intentional, creative, and original.
- Work meets all requirements of the assignment and is utilized to facilitate development of personal understanding.

LESSON 1.3: POETRY AND PERSPECTIVE

Writing Creatively to Imagine the Lives of Others

About this Lesson

In this lesson, students discuss the ways that poems they have read imagine the point of view of other creatures, how these creatures experience life, and what it would be like to live as one of these creatures. Then students write their own poems and short stories from the perspectives of other beings, imagining life as those creatures. In the process, students not only analyze and use poetic language, they also cultivate their capacity for empathetic imagination and forge a practice of relating to and valuing other beings.

Lesson Objectives

Students will:

- Gain knowledge of the features and techniques of poetry.
- Employ empathy and imagination to better understand the lives of others, including nonhuman creatures.
- Critically analyze poems.
- Compose original creative writing.
- Use language creatively to imagine and express the experiences and emotions of others, including nonhuman creatures.

Lesson Activities at a Glance

1. Review poems assigned in Lesson 1.1.
2. Conduct in-class discussion in groups and as a whole class.
3. Students complete a creative writing assignment.

Key Content Area Skills: analyzing poetry, writing poetry and creative prose.

Texts Used in this Lesson

There are no new assigned readings for this lesson; review texts from Lesson 1.1.

My Procedure

This lesson builds on the poems students read in Lesson 1.1. After completing that lesson, students have read these poems, written response papers reacting to them, and discussed them in class. I now ask students to review some of the poems, particularly those that explore the perspectives of nonhuman beings. We think specifically about "Sharks," "Don't Tell Me Not to Bother You," and "The Silence of Plants." I ask students to identify elements, passages, and techniques within these poems that express nonhuman perspectives. We talk about how the authors of these pieces have imagined those perspectives, and what details they've shared about them. We think about the sensory experience of being another type of creature other than human, about the motivations and desires of other creatures, about what these poems indicate we have in common and what is different.

At this point students have explored how these poems address the experiences of other beings, and many of them have started to identify with beings other than themselves, relating to the struggles and hopes of other creatures, imagining life in a different type of body, and de-centering their own thoughts in order to make room for empathetic consideration of others.

To further cultivate this empathetic imagining, I next give students a creative writing assignment, asking them to imagine the details of life as another type of creature – this may be a nonhuman animal like a cow or bird, or a plant like a tree, or an insect; any nonhuman living being. Here is the assignment:

Writing Assignment

Write a poem or short story from the perspective of a nonhuman being, describing an aspect of that being's experience. Think about how she/he/they experience daily life, who they interact with, what resources or conditions are important to them, who and what they love, what they fear, how they use their senses, and how they communicate. Consider their interactions with humans, if any, and how environmental degradation has affected them or will affect them.

Poems and short stories should demonstrate personal reflection and insight as well as detailed, empathetic consideration of the experiences of others, and should employ creative details, imagery, word choice, rhythm, style, and syntax to engage readers and evoke an emotional response.

The following is an example of a beautiful poem written by one of my students for this assignment.

Student Writing: Creative Writing Assignment

by Heather Harshbarger

"Caged"

> Tap, snapshot, leave. Repeat, refresh.
> Living, walking, breathing seems to mesh
> into daily routine through the glass
> as our paws tread over semi-synthetic grass.
> Life like the tomb, never slipping through
> the pack mentality I swear we once knew.
> Sniff the air before the putrid waft of sweat
> and urine and food puts us again in debt
> to the crypt keeper we never needed;
> our howls up to the moon, views impeded
> by a certain black darker than the stars,
> ears twitching from the flood of cars.
> Circle, nudge, attack. Repeat, retreat.
> This is our newer life, living, waking, defeat.
> Wait for the endless, moonless night,
> and struggle with every fickle fight
> to dominate the labyrinth of glass walls,
> how we long for the distant echoing calls
> that beckon; fading into the deafening cry
> of the crypt keeper that trained us to lie
> in the desolate crypt. Howl, yip, yelp,
> is the only way to seek unoffered help.
> The clock rings again, time to rise,
> I, the wolf, and put on the daily zoo disguise

This poem a great example of what to look for with this assignment: Heather writes from the perspective of a wolf in a zoo, vividly imagining the experience of living in captivity. Look for students to really try to get outside of their own experience, and to feel for and imagine the life of the creature whose viewpoint they are writing from.

As with Lesson 1.2, after students have written their assignments I ask if any are comfortable volunteering to share their pieces with the class, and we discuss any insights they gained from writing the assignment.

Products and Assessment

In this lesson students engage in class discussion in which they should be making specific reference to details from course readings. Through this process they should demonstrate careful reading, comprehension, and reflection. After this, students complete a creative writing assignment. This assignment should show creativity, thoughtful reflection, and empathy for other beings. Their participation in class discussion and their creative writing can be used to evaluate their achievement of lesson objectives.

I use the following grading criteria to evaluate student work for this lesson. I have, in different classes, formatted these criteria as a grading rubric or listed them in my course syllabus. How to format and apply the grading criteria will depend on individual context, school requirements, and teacher preference.

Grading Criteria

- Student demonstrates knowledge and critical analysis of the features and techniques of poetry.
- Student thoughtfully imagines the lives of others, including nonhuman creatures.
- Student demonstrates empathy in imagining the experiences, feelings, and desires of others.
- Student demonstrates thorough reading and comprehension of assigned course texts.
- Student demonstrates personal reflection, critical thought, and insight into course texts.
- Student's arguments are clear, well-developed, and documented with evidence from texts.
- Student's work demonstrates critical analysis of course topics, questions, and subject-matter.
- Style, usage, format, grammar, imagery, and presentation support meaning and are intentional, creative, and original.
- Work meets all requirements of the assignment and is utilized to facilitate development of personal understanding.

EFFECT OF THE LESSONS: EMPATHY AND RELATIONSHIP

> Love . . . is a habit or power of the mind, which grows and develops with use.
>
> —*Mary Midgley, qtd. in Peterson (48)*

They may not be highlighted as core skills in standardized learning objectives, but love, compassion, care, and empathy are indeed *skills* that can and must be cultivated. We can learn to imagine what someone else is feeling, to see the world from another's point of view. We can learn to feel for and with others, to value their subjective experiences as different from our own but equally deserving of acknowledgment, consideration, and respect. Consider the lessons in Unit 1 an exercise program for building these skills. As students read the assigned texts, discuss the perspectives presented by such diverse narrators, imagine the experiences of others, and think about important experiences they themselves have had interacting with other beings, they "decenter the self" (Peterson 2009, 35); they begin to see themselves as part of a larger picture, and they start to notice that being in relationship with people, nonhuman animals, plants, and landscapes is essential to emotional and physical well-being.

This may begin with simply recognizing the value that the natural world holds for humans. Students may start to reflect on all that the land does for them, including feeding them and providing the materials that form the objects they use in their lives. For example, here's a passage from a response paper in which a student who wishes to remain anonymous reflects on how dependent humans are on the natural world:

Student Writing: Response Papers

We are always connected to the land, whether we are breathing the air, eating the fruits of its soil, or simply lounging on the grass under the warming sun. All of the goods and services we want are merely a manufactured product made possible from the resources of the land.

Students may start to recognize emotional gifts they receive from the relationship with the land as well. They may discuss the benefits described by authors in Lesson 1.1, and they may suggest that connection to the natural world can offer inspiration, beauty, health, and joy. Here's an example of this from another student who wishes to remain anonymous, discussing the piece by Pam Houston assigned in Lesson 1.1:

Student Writing: Response Papers

I feel that this story highlights how the power of nature can completely change our emotions and feelings. Even something as simple as taking a walk around a neighborhood when we are having a busy day can provide a moment of clarity and peace, which leaves us feeling stronger and happier. . . . When we embrace it, nature often has a positive impact on our health and well-being. However, when we separate ourselves from nature, we often have a negative impact on the natural ecosystem. . . . If we embrace nature, we may reap fantastic benefits that improve our health and state of mind. However, if we choose to separate ourselves we lose this connection and do harm to the natural world at the same time.

As they move through the lessons in Unit 1, look for students to think like this about the value of relationship and connection between humans and the natural world. In order to truly feel such connection, however, students need more than to recognize that the land can benefit them. They need to feel compassion, empathy, and love for other beings. Wendell Berry has said, "to be grown up is to know that the self is not a place to live" (1995, 82). The cultivation of empathy is partially a process of learning and appreciating this truth. Berry suggests that we cannot, as individuals or as a culture, live solely inside ourselves, disregarding the existence and importance of other people, of nonhuman beings, and of the more-than-human world. Other authors have also suggested that humans *need* empathetic connection to the nonhuman world. Marc Bekoff echoes sentiments expressed by David Abram, saying "We are best understood in relationship with others . . . Animals are sources of wisdom, a way of knowing" (2003, 911).

This "way of knowing" can help us gain a wider perspective on the world, recognize the knowledge and needs of others, and discover the inherent benefits of forging close relationships with other beings, what Donna Haraway calls "'encounter value,' an addition to Karl Marx's categories of use and exchange value" (discussed in Peterson 2009, 35).

Many of the readings in Lesson 1.1, combined with the creative writing assignment in Lesson 1.3, are intended to challenge students to feel for others and to live beyond only their own interests. Here's an example of a student thinking about the interests of wolves, in response to a story by Terry Tempest Williams from Lesson 1.1:

Student Writing: Response Papers

by Chelsie Bateman

Wolves eat farmers' cattle and livestock and therefore the farmers kill them. But the farmers, and people in general, do not see that the wolves are doing this because people are intruding on their land, destroying their habitat, and making their prey disappear.

Empathy not only serves to increase compassion for others; by de-centering individual experiences and expanding the sphere of concern to the larger range of living beings with which we share the planet, empathy operates as a tool that can be applied to developing more thoughtful and inclusive approaches to the world. The lessons in Unit 1 should help students to stretch their sphere of concern and cultivate care for other beings. Here's an example of a student doing just this; in this excerpt student Jacob Rosenborough discusses a poem by Genny Lim from Lesson 1.1 in which the narrator releases a captive duck into the wild:

Student Writing: Response Papers

by Jacob Rosenborough

To me, someone with a "normal" perspective wouldn't think of animals as things that could be liberated, but Lim portrays the duck as a sort of slave that gets set free. It made me ask myself: do we enslave animals? And if so are we their rightful masters? A lot of people would probably think they're silly questions, and they may be, but humans have sort of assumed that we're smarter than other animals and can just sell their lives for $12.50; maybe we need to rethink our assumptions?

I hope that students participating in the lessons in Unit 1 will ask themselves exactly this sort of question – that they will look critically at their own assumptions, think deeply about the lives of other beings, and allow themselves to feel genuine affection and love for other animals, plants, and the entirety of the more-than-human world. As student Jennie Williams puts it, "humans must understand that we are part of the environment and we must care for nature as we care for ourselves."

UNIT 1 READINGS AND REFERENCES

Abbey, Edward. 1994. "The Serpents of Paradise." In *Reading the Environment*, edited by Melissa Walker, 1st ed., 51–56. New York: W. W. Norton.

Baca, Jimmy Santiago. 2001. "Ah Rain!" In *Poetry Like Bread*, edited by Martín Espada, 54. Willimantic: Curbstone Press.

Bekoff, Mark. 2003. "Minding Animals, Minding Earth: Old Brains, New Bottlenecks." *Zygon* 38 (4): 911–941.

Berry, Wendell. 1995. *Another Turn of the Crank: Essays*. Washington, DC: Counterpoint.

Brugnaro, Ferruccio. 2001. "Don't Tell Me Not to Bother You." In *Poetry Like Bread* edited by Martín Espada, 75. Willimantic: Curbstone Press.

de Nerval, Gérard. 1980. "Golden Lines." In *News of the Universe: Poems of Twofold Consciousness*, edited and translated by Robert Bly, 38. San Francisco: Sierra Club Books.

Dillard, Annie. 1994. "Living Like Weasels." In *Reading the Environment*, edited by Melissa Walker, 1st ed., 63–66. New York: W. W. Norton.

Houston, Pam. 1994. "A Blizzard Under Blue Sky." In *Reading the Environment*, edited by Melissa Walker, 1st ed., 57–62. New York: W. W. Norton.

Leopold, Aldo. 1970. *A Sand County Almanac: With Essays on Conservation from Round River*. 1st Ballantine Books ed. New York: Ballantine Books.

Lim, Genny. 2003. "Animal Liberation." In *From Totems to Hip-Hop: A Multicultural Anthology of Poetry Across the Americas, 1900–2002*, edited by Ishmael Reed, 34–36. New York: Thunder's Mouth Press.

Lorca, Federico García. 1980. "New York (Office and Attack)." In *News of the Universe: Poems of Twofold Consciousness*, edited and translated by Robert Bly, 110–112. San Francisco: Sierra Club Books.

Momaday, N. Scott. 1999. "The Man Made of Words." In *Our Land, Ourselves: Readings on People and Place*, edited by Peter Forbes, Ann Armbrecht, and Helen Whybrow, 2nd ed., 71–73. San Francisco: Trust for Public Land.

Peterson, Anna Lisa. 2009. *Everyday Ethics and Social Change: The Education of Desire*. New York: Columbia University Press.

Quintana, Leroy V. 2001. "Sharks." In *Poetry Like Bread*, edited by Martín Espada, 202–203. Willimantic: Curbstone Press.

Riedelsheimer, Thomas. 2002. *Rivers and Tides: Andy Goldsworthy Working with Time*. Documentary. Roxie Releasing.

Rilke, Rainer Maria. 1980. "I Live My Life." In *News of the Universe: Poems of Twofold Consciousness*, edited by Robert Bly, 76. San Francisco: Sierra Club Books.

Rilke, Rainer Maria. 1995. "Ah, Not to Be Cut off." In *Ahead of All Parting: The Selected Poetry and Prose of Rainer Maria Rilke*, 191. New York: Modern Library.

Shakur, Tupac. 1999. "The Rose That Grew from Concrete." In *The Rose That Grew from Concrete*, 3. New York: MTV/Pocket Books.

Snyder, Gary. 1994a. "The Call of the Wild." In *Reading the Environment*, edited by Melissa Walker, 1st ed., 71–74. New York: W. W. Norton.

Szymborska, Wisława. 1998a. "The Silence of Plants." In *Poems: New and Collected 1957–1997*, 269–270. San Diego: Harcourt.

Szymborska, Wisława. 1998b. "Among the Multitudes." In *Poems: New and Collected 1957–1997*, 267–68. San Diego: Harcourt.

Thoreau, Henry David. 1994. "Walking." In *Reading the Environment*, edited by Melissa Walker, 1st ed., 41–44. New York: W. W. Norton.

Williams, Terry Tempest. 1994. *An Unspoken Hunger: Stories from the Field*. New York: Pantheon Books.

Unit 2

LANGUAGE, MEDIA, AND WORLDVIEWS

Challenge hard-wired socialization

Everything you do to us will happen to you; we are your teachers, as you are ours. We are one lesson.

−Alice Walker, "Am I Blue"
Lesson 2.4

The earth was formed whole and continuous in the universe, without lines. The human mind arose in the universe needing lines, boundaries, distinctions. . . .

Between me and not-me there is surely a line, a clear distinction, or so it seems. But now that I look, where is that line?

This fresh apple, still cold and crisp from the morning dew, is not-me only until I eat it. When I eat, I eat the soil that nourished the apple. When I drink, the waters of the earth become me. With every breath I take in I draw in not-me and make it me. With every breath out, I exhale me into not-me.

If the air and the waters and the soils are poisoned, I am poisoned. Only if I believe the fiction of the lines more than the truth of the lineless planet will I poison the earth, which is myself.

−Donella Meadows, "Lines in the Mind" (53−54)
Lesson 2.2

The lessons in this unit are designed to help students recognize that the underlying belief systems held by a given culture are not simply "natural" or "given," but are created through language and culture and change over time and space.

By exploring diverse worldviews, comparing creation stories from around the globe, and studying the root metaphors we use to reason about the natural world, students can start to understand the patterns of thought that have led modern society toward destructive interactions with the larger world, and to explore alternative patterns of thought that do not support oppression or exploitation. This certainly doesn't mean that all aspects of the belief systems many students have grown up with are negative or destructive, or that they must be discarded – it just means that it is important to be aware of what effect these belief systems have on our attitudes and behaviors, and to be open to modifying the way we think, speak, and conceive of things in order to achieve more positive ends. It can be difficult to separate ourselves from our cultural norms and dominant beliefs, but students can come to realize that many mainstream assumptions – like what gender, race, or species is superior or inferior to others, or who has the "right" to use others for their own purposes – are not based in facts or inherent truths, but are formed by the people in power and passed on through culture. We don't have to hold on to these dominant assumptions; we can instead actively seek to adopt and communicate beliefs that encourage us to engage with the world in more equitable and sustainable ways.

Some key questions that the lessons in this unit should raise for students:

- What do you believe is the purpose of the natural world? What is the purpose of human beings?
- How do people acquire their beliefs?
- What messages do we get from our culture about our role in the world? How do we get these messages?
- What messages do we get in our culture about animals, plants, and the land? How do we get these messages?
- Does our culture believe some beings are superior to others? What leads us to believe this?

ethical

LESSON 2.1: WHAT WE BELIEVE

Examining Worldviews and Creation Stories

About this Lesson

In this lesson students are asked to think critically about beliefs held by society and how these beliefs came about. This includes gaining knowledge of different worldviews, comparing creation myths from diverse cultures, and asking challenging questions about how we see the world in modern society. Many of these beliefs are typically left unquestioned in mainstream culture, and often in the classroom as well. Conversations about such well-established assumptions can be difficult to start, but with the right scaffolding they prove extremely rewarding and valuable for students.

Lesson Objectives

Students will:

- Gain knowledge of established worldviews from diverse cultures.
- Gain knowledge of creation myths from diverse cultures.
- Critically analyze worldviews and creation myths and their influence on society.
- Think critically about the implications and effects of established worldviews and creation myths.
- Compose original analytical writing.

Lesson Activities at a Glance

1. Students read assigned texts on worldviews and creation myths from around the world.
2. Students write response papers outside of class.
3. Students answer questions in class discussion about the implications of varying worldviews.
4. In-class discussion about selected creation myths and their influence on culture.
5. Students summarize and analyze selected creation myths.

Key Content Area Skills: comparative cultural and historical analysis, critical analysis of societal beliefs and discourse, analytical writing.

Texts Used in this Lesson

James Sire, *The Universe Next Door: A Basic Worldview Catalog*

I have students read pages 13–20, then skim through 25–43, 50–56, 62–74, 85–88, 111–118, and 139–153.

Barbara C. Sproul, *Primal Myths: Creation Myths Around the World*, 49–59, 91–102, 122–126, 179–181, 199–200, 268–284, and 287–295

Lynn White, "The Historical Roots of Our Ecologic Crisis," *Science* 1203–1207

Optional: Ronald Wright, *A Short History of Progress*, 1–27

Note: Publication information for these readings is listed at the end of the unit. Sample passages are included in the full List of Readings at the end of the book.

My Procedure

I start this lesson by having students read the assigned selections. James Sire's piece describes varying worldviews and religious beliefs and the collection edited by Barbara Sproul recounts ancient creation myths from cultures around the world. These texts are meant to expose students to the idea that different cultures, religions, and groups of people have different views about the world, about what role humans should play in it, and about how we should treat other living beings and the land. After reading these texts, students write response papers outside of class. Here's part of a response paper in which a student comments on James Sire's text:

Student Writing: Response Papers

by Jennie Williams

[I]t was very thought provoking about the beliefs human beings possess. . . . To alter these perspectives causes us to see the world and ourselves from another point of view that would make the most subtle details of our lives come out in a whole new light.

In this text James Sire outlines several types of "worldviews" that he identifies and links with different religious traditions (including Deism, Naturalism, Nihilism, and Pantheistic Monism). When students come to class, after they have read the texts and written their response papers, I have them get into six groups. I assign each group one of the worldviews described in the reading, and for each one I ask them to think about the possible results of that worldview. I ask them to discuss how a person might live, what their goals might be, and how they might see the world and their role within it if they subscribe to that particular worldview. I post the following set of questions – it contains basic recaps of the six worldviews discussed in the reading, one for each group:

How might a person choose to live . . .

1. If the universe was created by an all-knowing God who has a plan?
2. If a powerful being created a clockwork universe which now runs toward a pre-determined end?

3. If the universe is simply matter and humans are complex machines, and everything operates on the laws of cause and effect?
4. If life has no purpose, meaning, or ultimate goal?
5. If the only meaning and value in life is that which humans make for themselves?
6. If each human being is one with the universe, and all are part of a spiritual whole?

Students discuss the possible outcomes of these worldviews in their groups and then share their conclusions with the class as a whole. What's key in this process is that they are beginning to actively understand the notion that worldviews are not universal or inherently true, but are a product of cultural influences and a source of cultural attitudes and behaviors. We often discuss the idea that most people may hold parts of more than one of these worldviews, and may be influenced by more than one cultural group or tradition. We highlight that none of these worldviews is intrinsically "bad," but that it's important to recognize how they influence our thinking, and to be aware that there are other ways of seeing the world.

Next we discuss and write about the creation myths students have read, from the collection by Barbara Sproul. We review one myth as a class, unpack what events take place in the myth, and then discuss what a culture that believes in this myth would think about their role in the world and how they should treat non-human animals, plants, and the land. I then have students pick two of the other myths they've read and write answers to the following questions. Sometimes I have them work on these questions with partners or in small groups, sometimes individually.

Writing Assignment

First write a very brief summary of the myth, describing the major events and figures.

- How was the earth created in this story?
- How were humans created?
- How were plants and nonhuman animals created?

Next think about what this story suggests about existence.

- Is there an all-powerful entity controlling what happens in this story? Is there a divine plan?
- What does the story suggest humans are here for?
- What does the story say our relationship to nonhumans should be?

Finally, how would a culture whose belief systems are based on this creation story think about the natural world?

My students are always interested to learn about creation stories from different cultures, and they do a great job of discussing the attitudes these myths might communicate within society. I'll share here an excerpt from a summary one student wrote of a creation story titled "When God Came to Earth," from the Nandi people of Africa. In the story, a man, an elephant, and thunder are living together on earth. Thunder is frightened of the man and retreats to the heavens. The man is pleased that the thunder left, and he fashions a poisoned arrow that he uses to kill the elephant. He then becomes the most powerful being on earth. Here's what my student writes of this myth:

Student Writing: Creation Myth Analysis

by Christina Malliakos

This story does not say how earth or animals were created but it does go into why man is so powerful. Thunder and the elephant were talking about man and thunder was scared because humans have the ability to turn in their sleep. Elephant laughed and then was shot by a poisonous arrow by the human. Thunder then said that "I shall not take you [to heavens], for when I warned you that the man was bad, you laughed and said he was small" ("When God Came to Earth," paragraph 7). Then the elephant died and it shows that man is harmful to nature but will ultimately have control over nature even if it is bad or good.

Students should identify the assumptions within each myth, such as the assumption in this myth that humans always had and will always have control over the earth from the very beginning. Additionally, look for students to not only highlight what conclusions the story might lead people to draw, but also perhaps to explore possible alternatives that are missing from the myth. For example, some students have pointed out that, by killing the elephant, humankind misses out on the chance to have a positive relationship with the elephant, and with other species of animals in general.

The selection of creation myths we read also includes Genesis 1 and 2, and students are quick to notice the implications of this creation story when compared to the others they read. Here's one example from an anonymous student:

Student Writing: Creation Myth Analysis

In the book of Genesis, God creates everything in a series of seven days. First comes light, then the earth, vegetation from the earth, the sun and the moon, animals on the land and in the seas, the creation of man in God's own image, and finally a day of rest. While this is the basic summary of Genesis, there are some very important implications of this creation myth. In the story, everything is created by God. Were it not for God's divine power, nothing would be here. Therefore, believers of this story are very reverent to God and devoutly believe in a higher power. Also, God states that man is to be master of the plants and animals of the earth. This has powerful consequences as it does not state that humans are to be stewards of the environment. Rather, it suggests that everything on earth is for our taking. Adhering to this belief can lead to unsustainable practices which do a great deal of harm to the natural world.

In their analyses of Genesis 1 and 2, students swiftly notice the assumption that the earth has been given to humans to use. However, many also point out interesting differences between the two versions of the story. For example, they may note that Genesis 1 states that God gave humans "dominion over the earth," but also states that God tells man, "I have given you every plant yielding seed which is upon the face of all the earth, and every tree with seed in its fruit; you shall have them for food" (Sproul 124), notably not listing animals as food for humans. Meanwhile Genesis 2 says that God created other animals as "helpers" for man, but not finding one fit to serve this role, God created woman to serve as man's helper. There are interesting implications to unpack here.

Products and Assessment

In this lesson students write summaries and analyses of creation myths, as well as response papers in which they react to the concepts in the assigned readings. In addition to their participation in class discussion, these written products can be used to evaluate students' achievement of the lesson objectives. Students' creation myth analyses should demonstrate understanding of the events of the stories and critical insight into the social implications of believing the assumptions embedded in the myths.

I use the following grading criteria to evaluate student work for this lesson. I have, in different classes, formatted these criteria as a grading rubric or listed them in my course syllabus. How to format and apply the grading criteria will depend on individual context, school requirements, and teacher preference.

Grading Criteria

- Student compares and analyzes diverse worldviews, and demonstrates critical thinking about the implications and social construction of cultural belief systems.
- Student demonstrates critical thinking about the role and implications of creation narratives from diverse cultures.
- Work demonstrates thorough reading and comprehension of assigned course texts.
- Work demonstrates personal reflection, critical thought, and insight into course texts.
- Arguments are clear, well-developed, and documented with evidence from texts.
- Work demonstrates critical analysis of course topics, questions, and subject-matter.
- Style, usage, format, grammar, imagery, and presentation support meaning and are intentional, creative, and original.
- Work meets all requirements of the assignment and is utilized to facilitate development of personal understanding.

LESSON 2.2: MASTERS, STEWARDS, FAMILY

Cultural Metaphors for the Natural World

About this Lesson

In this lesson students are asked to continue to think critically about beliefs held by society and how these beliefs came about. Students learn about the concept of "conceptual metaphor," and how societies use metaphors to understand their world. They critically examine several common metaphors used in U.S.-American society for thinking about the natural world, and they consider the consequences of adopting each of these metaphors. Students then imagine alternate metaphors they believe might lead to more positive behaviors.

Lesson Objectives

Students will:

- Gain knowledge of conceptual metaphors and their function in cultural discourse.
- Critically analyze examples of conceptual metaphors.
- Think critically about the implications and effects of established cultural metaphors about the natural world.
- Think creatively about alternative conceptual metaphors and their possible implications.
- Compose original analytical writing.

Lesson Activities at a Glance

1. Students read about conceptual metaphor and write response papers outside of class.
2. In-class discussion about conceptual metaphor and specific examples.
3. Students analyze the effects of selected conceptual metaphors and propose alternate metaphors for thinking differently about the natural world.

Key Content Area Skills: comparative cultural and historical analysis, critical analysis of societal beliefs and discourse, analytical writing.

Texts Used in This Lesson

Texts on Conceptual Metaphor

George Lakoff and Mark Johnson, *Metaphors We Live By*, 3–13, 22–24

Rebecca Martusewicz, Jeff Edmundson, and John Lupinacci, *EcoJustice Education*:
"Metaphors and the Construction of Thought," 59–62
"Discourses of Modernity," 66–68

Passages to Analyze as Examples

Donella H. Meadows, "Lines in the Mind," in *Our Land, Ourselves*, 53–55

Passages from *Reading the Environment*, edited by Melissa Walker:
"What is Biodiversity and Why Should We Care About It" by Donella Meadows, 149–151
"Storm Over the Amazon" by E.O. Wilson, 151–161
"Mites, Moths, Bats, and Mosquitoes" by Sue Hubbell, 161–164

Passages from *Environmental Discourse and Practice* edited by Lisa Benton and John Rennie Short:
"How Can One Sell the Air?" by Chief Seattle, 12–13
"The Hoop of the World" by Black Elk, 257
"Essay on American Scenery (1835)" by Thomas Cole, 87–90
"A Voice for Wilderness (1901)" by John Muir, 102–104
"Conservation, Protection, Reclamation, and Irrigation (1901)" by Theodore Roosevelt, 110–113
"The Obligation to Endure" by Rachel Carson, 126–128
"Message to Congress (1970)" by Richard Nixon, 132–139
"Confessions of an Eco-Warrior" by Dave Foreman, 198–200
"Ecofeminism" by Carolyn Merchant, 209–213

Wisława Szymborska, "Water," in *Poems: New and Collected 1957–1997*, 58–59

Juan Felipe Herrera, "Earth Chorus," in *From Totems to Hip-Hop: A Multicultural Anthology of Poetry Across the Americas, 1900–2002*, 30–32

May Swenson, "Weather," in *From Totems to Hip-Hop: A Multicultural Anthology of Poetry Across the Americas, 1900–2002*, 52–53

Aldo Leopold, *A Sand County Almanac*:
"Thinking Like a Mountain," 137–141
"Prairie Birthday," 47–54
"The Land Pyramid," 251–258

Naomi Klein, "A Hole in the World," in *The Nation*

Optional: Bill McKibben, *The End of Nature*, 47–61, 77–91

Optional: Wendell Berry, *Another Turn of the Crank*, 64–77

Note: Publication information for these readings is listed at the end of the unit. Sample passages are included in the full List of Readings at the end of the book.

My Procedure

In this lesson I introduce students to the idea of "conceptual metaphor," described by George Lakoff and Mark Johnson. The idea here is that metaphors aren't just a literary tool, they're how we think about the world, by using our knowledge about one subject to reason about something else. Lakoff and Johnson give some common examples of conceptual metaphors in U.S. society, like the metaphor "argument is war" or "time is money." Whether we directly articulate it or not, we may find ourselves reasoning about argument the way we would reason about war, in terms of "attacking," "strategizing," "winning," and "losing." And we often think about time the way we would about money, as a limited commodity, something to save, give, take, or lose (4–8). When we think about something using a conceptual metaphor, we apply the assumptions and logic that would apply to one thing onto something else, and we often don't realize we're doing it. Through this process of comparison, our attention will be directed toward certain conclusions and other considerations will be obscured from our view. The passage from Martusewicz, Edmundson, and Lupinacci takes up this notion of conceptual metaphor and starts to explain its implications for our thinking about the natural world.

First students read the assigned passages by Lakoff and Johnson and by Martusewicz, Edmundson, and Lupinacci outside of class, and then complete response papers. Some struggle with the idea of conceptual metaphor a bit, but reading each other's response papers is a great start, as the students who have gotten a strong grasp on the material can help those who haven't. Here's an example of what students say about Lakoff and Johnson in their response papers:

Student Writing: Response Papers

by Jennie Williams

I was . . . intrigued and fascinated by their content because they were concentrated on the natural philosophies and unnoticed tendencies of our lives. In "Metaphors We Live By," I frequently had to stop while reading to speak out loud some of these phrases categorized by argument is war. "I've never won an argument . . . He shot down all of my arguments." . . . Obviously no one pulled out a literal gun and shot the argument, however our language uses words of force and brutality to express arguments. I was stunned when the passage made me, "imagine a culture where an argument is viewed as a dance . . . In such a culture, people would view arguments differently" (5). Arguments would not have such a negative and aggressive connotation if they were viewed as more stylistic and elegant like a dance. It may be easier and less emotional to present our points and differences in views if we argued in a different sense of metaphor. Since I have read "Metaphors We Live By," I have been examining the words I say in depth to see if I have been using metaphorical terms to describe what message I try to convey. It was incredible to me how many times I stopped to think about what I've said!

When students come to class we discuss this reading further and review the idea of conceptual metaphor. Once students are fairly clear on this idea, we next discuss how conceptual metaphor applies to the way our culture thinks about the natural world. How do we each think about the earth? What metaphors do we use, knowingly or not, to reason about the natural world and to draw conclusions about how we should treat it? In class I offer examples of several metaphors for the natural world commonly used in U.S. culture, and we discuss the implications of each. (I use a slideshow for this discussion that can be found in the online resource for this book.)

The first metaphor I discuss is "the natural world is a stockpile of raw materials." If we see the world as a collection of objects for our use, we are likely to think that our role in the world is to extract those "resources." We'll think about how to use and distribute the materials, and perhaps about how to avoid using them too quickly. But this metaphor obscures the subjectivity and needs of other beings – within this framework, the earth and all other beings that inhabit the earth are presented as an inanimate collection of materials, a commodity.

Another example I use is "the natural world is a garden." Seeing the world as a garden can highlight issues of care, encouraging us to "tend" to the earth. However, it can also imply human dominance, since this metaphor has the potential to make us believe that it is our right and responsibility to arrange other living beings and decide which can grow and which cannot.

I also discuss the metaphor "the natural world is family." If we see the world as a family, we may think of our own welfare and the welfare of other living creatures as linked. We may see ourselves of part of the natural world, and perhaps focus on interdependence.

Each of the views described above offers a different perspective for relating to the world, implies a different set of assumptions about ourselves and about other beings, highlights a different set of questions, and obscures different considerations.

I discuss these common metaphors with my students, and for each one I ask them to consider the results of seeing the world this way, and to think about what sorts of conclusions each metaphor directs our attention toward and what considerations it obscures, hides, or leaves out.

Class discussion of these metaphors is always energetic and excited, and my students quickly raise terrific insights about the implications of each metaphor. They notice pros and cons of each, and often start to identify alternatives that they prefer. Many conclude that their favorite alternative is "the natural world is a community," a metaphor that highlights mutual well-being and partnership.

Throughout this discussion, students should be considering the cultural motivations that lead people to adopt and maintain their values and behaviors and how these background patterns of thought shape attitudes, often without our knowledge. They should also be starting to think about alternative views, values, and approaches to the world that might lead to more just and healthy behaviors.

After discussing these metaphors, I have students practice identifying metaphors at work in literary passages – these passages are the other texts in the list of readings above. They include pieces by classic environmental writers and by contemporary authors and poets. I typically assign these readings in advance for students to read outside of class, and during class I assign one passage each to students, usually in small groups. For each passage, I ask students to identify what metaphors about the natural world the author is using (whether those metaphors are directly stated or below the surface), and what other beliefs and worldviews the author is expressing. Groups discuss and identify the metaphors and worldviews they see at work in the passage, and then they share their conclusions with the class.

Students note a range of metaphors at work in these passages during this discussion; many have also noted some of these metaphors and thought patterns in their response papers prior to discussing them in class. For example, speaking of Naomi Klein's article "A Hole in the World," in which Klein discusses several interesting metaphors, one anonymous student writes:

Student Writing: Response Papers

It seems as if oil executives and some politicians on both sides of the aisle view nature as something to be controlled and taken advantage of.

Another student comments on the view that humans are "masters" and "conquerors" of the planet:

Student Writing: Response Papers

by Jacob Rosenborough

[I]t is apparent that the "master" mindset of humans "ruling over" the Earth hasn't gotten us well off in the environmental field of things. . . . we treat the Earth as something that needs to be conquered, and going to war with literally everything that constitutes our existence is a pretty bad idea.

And in one more example, a student discusses the metaphor of nature as a "web," expressed in a piece titled "Mites, Moths, Bats, and Mosquitoes" by Sue Hubbell:

Student Writing: Response Papers

by Michelle Ott

After a great deal of pondering, I was finally able to realize the deeper meaning behind a detailed story that intertwines the lives of the mites, moths, bats, and mosquitoes. Hubbell has offered a story to emphasize the fact that there are a countless number of relationships that exist amongst people, animals, insects, and the natural world. I found it to be rather interesting that a Moth Ear Mite is ultimately in a relationship with a human. The mite makes the moth an easy target. The bat catches the moth and provides him with the nourishment he needs to be able to catch the mosquito. The bat is protecting the human from the mosquito. After being engaged in such deep thought, I came to realize the world is similar to a web. As Hubbell concluded, "So there we are out under the oak trees in the dim light – the mites, the moths, the bats, the mosquitoes, and me. We are a text of suitability one for another, and that text is as good as any I know by which to drink my coffee and watch the dawn" (164). The entire world is connected in some way, shape, or form. Whether humanity is aware of it or not, we are inevitably . . . connected to the natural world. . . . Hubbell was able to enjoy her coffee because her friends were swarming around her.

Finally, after discussing these passages, I ask students to think about the views, beliefs, and metaphors that they each use themselves, and which ones they think would lead to the most positive and sustainable behavior. They often raise valuable insights about their own viewpoints and those they have grown up with in our culture.

I continue to link this subject to other topics throughout the curriculum, asking students to analyze films on environmental topics and discuss the worldviews and attitudes toward the natural world they see in the films, and raising the subject in reference to other readings later in the course. Students continue to raise the subject themselves, as well; in future lessons in this book you may notice students referencing metaphors they had previously discussed during this lesson.

Products and Assessment

In this lesson students write response papers in which they react to the concepts in the assigned readings. They also answer questions about what metaphors are at work in sample passages – this can be submitted in writing or shared orally in class. In addition to their participation in class discussion, these products can be used to evaluate students' achievement of the lesson objectives.

I use the following grading criteria to evaluate student work for this lesson. I have, in different classes, formatted these criteria as a grading rubric or listed them in my course syllabus. How to format and apply the grading criteria will depend on individual context, school requirements, and teacher preference.

Grading Criteria

- Student demonstrates an understanding of the idea of conceptual metaphor and how it applies to thinking about the natural world.
- Student demonstrates critical analysis of specific conceptual metaphors.
- Student demonstrates critical thinking about the implications of dominant cultural metaphors.
- Student demonstrates creativity and imagination in proposing alternative conceptual metaphors.
- Work demonstrates thorough reading and comprehension of assigned course texts.
- Work demonstrates personal reflection, critical thought, and insight into course texts.
- Arguments are clear, well-developed, and documented with evidence from texts.
- Work demonstrates critical analysis of course topics, questions, and subject-matter.
- Style, usage, format, grammar, imagery, and presentation support meaning and are intentional, creative, and original.
- Work meets all requirements of the assignment and is utilized to facilitate development of personal understanding.

LESSON 2.3: LANGUAGE AND DOMINATION

How What We Say Shapes What We Believe about Others

About this Lesson

In this lesson students are shown examples of how word choice and discourse can influence attitudes. They explore the ways that standard language choices communicate certain assumptions about nonhuman animals and the land, and discuss how exploitation of human and nonhuman groups can be perpetuated or justified through language.

Lesson Objectives

Students will:

- Gain knowledge of concepts from discourse studies and ecolinguistics.
- Critically analyze language choices and how they influence attitudes and behavior.
- Think critically about standard language usage as it relates to oppressed people, nonhuman animals, and the land.
- Compose original analytical writing.

Lesson Activities at a Glance

1. Students read about language and discourse, and their role in the domination of human and nonhuman groups.
2. Students write response papers outside of class.
3. In-class discussion about language use and examples of its influence; students analyze the effects of selected words and phrases discussed in the readings.

Key Content Area Skills: critical analysis of language, critical analysis of societal beliefs and discourse, analytical writing.

Texts Used in this Lesson

James Paul Gee, *Social Linguistics and Literacies: Ideology in Discourses*, 6–15

Joan Dunayer, *Animal Equality: Language and Liberation*, 1–20, 179–201

Robert Cox, *Environmental Communication and the Public Sphere*, 23–28, 58–70, 152–157, 163–165, 174–179

Aldo Leopold, *A Sand County Almanac*: "Pines Above the Snow," 86–93

Fritjof Capra, *The Hidden Connections*, 54–64

Tania Soussan, "Scientist: Prairie Dogs Have Own Language," *redOrbit*

James Honeyborne, "Elephants Really Do Grieve Like Us: They Shed Tears and Even Try to 'Bury' Their Dead – A Leading Wildlife Film-Maker Reveals How the Animals Are Like Us," Mail Online

Jennifer Viegas, "Chickens Worry about the Future," *ABC Science*

Optional: Cathy Glenn, "Constructing Consumables and Consent: A Critical Analysis of Factory Farm Industry Discourse," *Journal of Communication Inquiry*, 63–81

Optional: Peter Mühlhäusler, *Language of Environment, Environment of Language: A Course in Ecolinguistics*, 15–26

Note: Publication information for these readings is listed at the end of the unit. Sample passages are included in the full List of Readings at the end of the book.

My Procedure

I start this lesson having students read a selection of passages that explore how language influences our attitudes, and specifically how it influences our thinking about nonhuman animals and the natural environment. Gee provides a general introduction to the concept of discourse studies and the fact that language is not a concrete object, but something that changes with culture. Dunayer discusses the different vocabulary and linguistic practices commonly used when talking about nonhuman animals versus talking about humans – for example, using the pronoun "it" rather than "she" or "he" to talk about other living beings, or excluding humans from the category of "animal" by using that term only to refer to nonhumans. She argues that our language use implies human separateness and superiority, leading us to assume that we're more different from other animals than we actually are, and helping authorize our belief that humans are smarter and better than other creatures. Her piece reminds us that language can direct our attitudes and attention, and by choosing certain words over others we might be led to feel differently about a particular idea. Students read these assigned passages for this lesson and write a response paper prior to class, articulating their initial reactions and questions. Here's an example of what students have to say about the passage by Gee:

Student Writing: Response Papers

This week's reading materials strongly stated how language is a valuable tool that shapes human's beliefs, attitudes, opinions, and perspectives. The passage Social Linguistics and Literacies by Gee really opened my mind to the semantics and diverse connotations a single word possesses. Gee argues that words do not have a fixed definition, vary on context, and are cultural models that can change (15).

Below is part of a response paper written about Joan Dunayer's piece.

Student Writing: Response Papers

by Michelle Ott

Throughout history, different cultures and ethnic groups have faced a great deal of discrimination; however, humanity now desires to see everyone as the same. If humanity envisions equality, why don't we treat every living and breathing creature the same? Why do we feel as though we are superior to other animals? I became increasingly disgusted with humanity as I read through the piece, Animal Equality: Language and Liberation, by Joan Dunayer. Every member of society should be degraded as a speciesist. We fail to give nonhuman animals equal consideration, respect, and rights. Dunayer opened my eyes as I realized that I am also guilty of such a crime. I don't refer to animals as he or she. I use the pronoun, it, when referencing the living and breathing nonhuman animal. It is evident through everyday language that humanity thinks it is higher and better than every other species. Dunayer makes the point that we capitalize and give a plural sense to the word, Man. We do not refer to any other species in this sense. Why does Man get capitalized? Do we consider ourselves above all and to be like God? Dunayer makes the assertion, "Directly or indirectly, most humans routinely, knowingly participate in unnecessary harm to nonhumans, including their torture, imprisonment, enslavement, and mass murder" (4). After realizing the truth behind Dunayer's words, I felt ashamed and disappointed in mankind. When a human is killed, the murderer is thrown in jail. Why is it that when the victim is a nonhuman animal the murderer is not punished in the same way? Why don't we use the same words to describe animal cruelty as we do to describe crimes? As I continued to read Dunayer's work and learned more about speciesism, I became irritated and angered by humanity.

Some of the readings assigned for this lesson can be a bit abstract, and the concept that our language influences attitudes that we normally don't question as a society is one that can be tough for students to wrap their minds around. For this reason, I usually follow the readings with a review in class of some of the concepts and examples presented in the texts. (See the slideshow I use for this lesson in the online resource.)

Having reviewed these concepts, I ask students to share their reactions to what they have read. There is usually a wide mix of opinions, from those who feel these ideas give them new insights to those who are skeptical, or feel challenged. There are usually some students who feel particularly defensive about the assertion that humans are not inherently superior to other species. The common assumption of human superiority is challenged by readings that point out that other species have language, grieving practices, friendships, and many other traits that some have claimed only belong to humans. This assertion will be deconstructed further in the next lesson as well; for now it's important for students to recognize that, whether they believe a given assumption or not, that assumption may be culturally constructed and is often communicated to us through language in ways we're not even aware of. I strongly recommend teaching this lesson and the following one in close succession, to continue the exploration of these culturally constructed beliefs of superiority.

I conclude by reviewing some of Dunayer's recommendations for "relanguaging," making the conscious choice to use language in ways that communicate respect and equality toward other beings and counteract social norms that make us assume hierarchies of superiority and importance. Some students are skeptical about implementing any change in linguistic habits. Others start implementing some of Dunayer's recommendations in their writing and their in-class comments. As one of my students comments:

Student Writing: Response Papers

by Heather Harshbarger

There are the imaginary lines and boundaries that are recurring in both nature and society that have been set up by humans as a separation. It's one way I believe speciesism is occurring in the world. As long as people continue to group things into "us" and "them," there won't truly be dissolution of the barrier between the animals and "us," or the regrouping of ourselves into the categories of animals. Why else would we call each other animals over something that seems lower than what humans do, or rather, why are humans considered the best species? We hear often about how other animals are "human-like" in their intelligence, as if human is the best standard there is. No one ever said we truly are the best species.

Products and Assessment

In this lesson students write response papers in which they react to the concepts in the assigned readings. In addition to their participation in class discussion, these written products can be used to evaluate students' achievement of the lesson objectives.

I use the following grading criteria to evaluate student work for this lesson. I have, in different classes, formatted these criteria as a grading rubric or listed them in my course syllabus. How to format and apply the grading criteria will depend on individual context, school requirements, and teacher preference.

Grading Criteria

- Student demonstrates an understanding of the idea of discourse, and how our language use can influence attitudes in ways we are not always aware of.

- Student demonstrates critical analysis of specific words and phrases and their impact on our attitudes about other beings and the natural world.
- Student demonstrates critical analysis of the arguments made by authors, how they play out in their lived experiences, and their relevance to social and environmental justice.
- Work demonstrates thorough reading and comprehension of assigned course texts.
- Work demonstrates personal reflection, critical thought, and insight into course texts.
- Arguments are clear, well-developed, and documented with evidence from texts.
- Work demonstrates critical analysis of course topics, questions, and subject-matter.
- Style, usage, format, grammar, imagery, and presentation support meaning and are intentional, creative, and original.
- Work meets all requirements of the assignment and is utilized to facilitate development of personal understanding.

LESSON 2.4: GENDER, RACE, SPECIES

Mutual Oppression

About this Lesson

In this lesson students are introduced to ideas from the field of ecofeminism, and they consider the ways that the oppression of human groups and of nonhuman animals reinforce one another. They look at historical examples of hierarchical thinking that values some people above others and values humans above other animals. They also read examples of primary historical writing that justifies oppression of certain human groups to see these "logics of domination" at work.

Lesson Objectives

Students will:

- Gain knowledge of concepts from ecofeminism.
- Gain knowledge of historical justifications for the oppression of human groups.
- Relate the domination of humans to the domination of nonhuman animals and the land.
- Critically analyze these forms of domination as mutually reinforcing.
- Compose original analytical writing.

Lesson Activities at a Glance

1. Students read about ecofeminism and the historical and modern domination of human and nonhuman groups.
2. Students write response papers outside of class.
3. In-class discussion about forms of domination and their relationship to each other.
4. Students read historical and contemporary writings and identify language features in the passages that evoke hierarchical thinking about the value of certain beings over others.

Key Content Area Skills: comparative cultural and historical analysis, critical analysis of societal beliefs and discourse, analytical writing.

Texts Used in this Lesson

Karen Warren, *Ecofeminist Philosophy*, 21–38

Rebecca Martusewicz, Jeff Edmundson, and John Lupinacci, *EcoJustice Education*: "Language, Dualism, and Hierarchized Thinking," 57–58

Alice Walker, "Am I Blue," *The Westcoast Post*

S. Snyder, "The Great Chain of Being." Grand View University

Historical Passages to Analyze in Class

Thomas Jefferson, "Notes on the State of Virginia," *Electronic Text Center, University of Virginia Library*, excerpt from Query 14, pages 264–267

Aristotle, "History of Animals," Book IX, Part 1, Paragraphs 5–7

Aristotle, "Politics," Part XII

J. B. Sanford, "Arguments Against Women's Suffrage, 1911" (see online resource)

"Vote NO On Woman Suffrage," National Association Opposed to Woman Suffrage (see online resource)

Note: Publication information for these readings is listed at the end of the unit. Sample passages are included in the full List of Readings at the end of the book.

My Procedure

I begin by assigning the readings listed above. The texts by Warren and Martusewicz, Edmundson, and Lupinacci introduce students to ecofeminism and the idea of "value hierarchies" and "logics of domination." These passages raise the notion that oppression of human groups and nonhuman beings is mutually reinforcing – that oppressed groups such as women and people of color have throughout history often been presented as inferior to the dominant groups in society, as have nonhuman animals and the land, and that these oppressed groups have been linked to one another to reinforce the view that they are inferior. They also raise the idea that those in power have throughout history claimed inherent superiority over exploited groups in order to gain or maintain their position of power, and that groups have been categorized into dualistic relationships in which one is superior to another. It's important for students to start to understand that discrimination is a cultural construction, and that ideas about who is more or less valuable change over time and often serve the needs of those in power. By seeing the historical and cultural forces at work, they may start to see what forces are still at work today. These related forms of discrimination are then beautifully illustrated by Alice Walker's "Am I Blue." After students read these assigned texts, they write response papers prior to class. Here's an example of a student discussing Warren:

Student Writing: Response Papers

by Christina Malliakos

One aspect that this passage talks about is how there is an aspect of hierarchy in the Christian religion. Man is at the top of this chain and then women come next because women were created from man. Then after humans are the animals and then the rest of nature. It shows that some cultures believe that humans have a higher ground on nature because we possess the ability to be logical. However, I think that this passage was trying to say that there really is no hierarchy but everyone is equal. This means that men and women are equals and that nature is just as important as humans to life on the planet Earth.

After students have read the texts and completed their response papers, I begin by reviewing some key concepts and examples from ecofeminism (see the online resource for the slideshow I use in my class). After this review I continue into a full class discussion, asking students to share their initial reactions to the concepts from these readings. Students often acknowledge gender, race, species, and other forms of discrimination but express doubt that some of them will ever change, especially in regards to other species. Some students can continue to feel personally challenged by these topics, as they may have in Lesson 2.3, so be prepared for students to get a bit defensive; looking at these forms of oppression means recognizing deeply rooted injustices in mainstream contemporary society, and for those students who identify strongly with mainstream society, any realization of injustices it perpetrates can feel like an attack on themselves and their way of life. Be careful to frame these discussions to make it clear that many of these processes of oppression happen below the surface, and people who use these sorts of discriminatory language often don't realize it, since it is so well integrated into society. Remind students that this isn't about attacking any individual person's behavior, it's about recognizing how our society operates so that we can consider more positive alternatives that make for a happier and more just society for everyone. Allow space for students to voice their doubts and work through any initial hesitation in seriously considering these arguments.

Next I introduce the notion of the "Great Chain of Being" and discuss how hierarchies of superiority have influenced beliefs for centuries. Throughout this discussion, remind students to consider who benefits from a particular belief system; for example, if women are considered inferior to men, who benefits and in what ways? If nonhuman animals are considered inferior to humans, who benefits and in what ways?

After discussing the "Great Chain of Being," I put students into small groups, and give each group a historical passage. These passages are examples in which the authors justify believing that certain groups of humans are inferior to other groups. The passages give students a chance to see these hierarchies in action, and realize how much damage such discriminatory beliefs can do. I ask students to discuss these passages in their groups, and to analyze the language that is used in them. Draw on discussions from Lesson 2.2 and Lesson 2.3 here to think about the specific words, phrases, and metaphors that are used. I ask students to identify examples of word use that reinforces attitudes of superiority or inferiority. Then I have students share their analyses with the rest of the class.

After each group has shared with the class, I ask students to return to their groups and come up with at least one way to counteract the "logics of domination" exemplified by the passages they read and by the modern examples of oppression described by ecofeminist scholars like Karen Warren. I have them consider "relanguaging" strategies, as discussed in Lesson 2.3, as well as any other approaches. Finally I have each group share these with the class.

Students should clearly see through this process that arguments that have been made justifying the exploitation of certain groups look very similar throughout history, though which groups they are applied to changes. For example, here's a passage in which a student points out that women's rights which we have started to see as normal today were not always thought of this way, and that perhaps this sort of change might apply to other species in the future:

Student Writing: Response Papers

by Jacob Rosenborough

[L]ooking back on it, we think that women's rights are a natural given and it shouldn't have been any other way before that. If this applies to animal rights in the future than I think the World will have made a step down the right path.

And here's a similar comment from an anonymous student on historical shifts in our attitudes toward exploited human groups and how such a shift could occur in our thinking about nonhuman groups as well:

Student Writing: Response Papers

Just because animals do not have rights now does not mean that they should not ever have rights. Just like slaves, women, and children were once viewed as just property and did not have any rights according to the law, they eventually gained rights.

Products and Assessment

In this lesson students write response papers in which they react to the concepts in the assigned readings. They also analyze materials in class and participate in class discussion. Their participation in this discussion and analysis as well as their written response papers can be used to evaluate their achievement of the lesson objectives.

I use the following grading criteria to evaluate student work for this lesson. I have, in different classes, formatted these criteria as a grading rubric or listed them in my course syllabus. How to format and apply the grading criteria will depend on individual context, school requirements, and teacher preference.

Grading Criteria

- Student demonstrates an understanding of concepts from ecofeminism, including the notion of mutually reinforcing oppression.
- Student demonstrates an understanding of the concept of "logics of domination" and how such logics have operated at past points in history and today.
- Student analyzes relationships between the domination of human groups and the domination of nonhumans and the land.

- Student demonstrates critical analysis of specific words and phrases and their impact on our attitudes about other humans, nonhuman beings, and the natural world.
- Student demonstrates critical analysis of the arguments made by authors, how they play out in their lived experiences, and their relevance to social and environmental justice.
- Work demonstrates thorough reading and comprehension of assigned course texts.
- Work demonstrates personal reflection, critical thought, and insight into course texts.
- Arguments are clear, well-developed, and documented with evidence from texts.
- Work demonstrates critical analysis of course topics, questions, and subject-matter.
- Style, usage, format, grammar, imagery, and presentation support meaning and are intentional, creative, and original.
- Work meets all requirements of the assignment and is utilized to facilitate development of personal understanding.

LESSON 2.5: MEDIA AND PUBLIC MESSAGES ABOUT THE ENVIRONMENT

Critical Media Literacy for EcoJustice

About this Lesson

In this lesson students explore the role of media in shaping public perception. They review rhetorical and visual techniques for directing an audience's attention and opinions, and they investigate potential forms of bias within news media and the reasons for these. Students then view and analyze examples of media products that convey distinct impressions about environmental issues, and identify the techniques used in those examples.

Lesson Objectives

Students will:

- Gain knowledge of rhetorical and visual techniques used by media outlets to frame information and influence opinion.
- Gain knowledge of possible forms of bias in news media.
- Critically analyze media products to determine the techniques, messages, and motivations at work.
- Compose original analytical writing.

Lesson Activities at a Glance

1. Students read about media, including advertising and news coverage of environmental issues.
2. Students write response papers outside of class.
3. In-class review of rhetorical and visual techniques and how they influence perception.
4. In-class analysis of examples of media products.
5. Optional: Follow in-class discussion with individual or groups analysis of further examples of media products.

Key Content Area Skills: critical analysis of language, rhetoric, and visual communication, critical analysis of media products, analytical writing.

Texts Used in this Lesson

Julia Corbett, "A Faint Green Sell: Advertising and the Natural World," in *Enviropop*, 141–160

Bill McKibben, *The Age of Missing Information*, 8–28

Shirley Biagi, *Media/Impact*, Ch. 10: Advertising: Motivating Customers, 213–230

"Media/Political Bias," *Rhetorica*. http://rhetorica.net/bias.htm

Note: Publication information for these readings is listed at the end of the unit. Sample passages are included in the full List of Readings at the end of the book.

My Procedure

To start this lesson, students read the assigned materials and write response papers outside of class. The readings for this lesson cover critiques of how images of the natural world are used in advertising, the impact of heavy television watching on our knowledge of the land, types of bias in news media, and what needs advertisers try to appeal to in order to entice audiences.

The pieces in this lesson present many facets of the media's influence on society, and my students usually respond with a good degree of criticality. Here's an anonymous student discussing Corbett's piece in his response paper:

Student Writing: Response Papers

I think that is it terrible when advertisers and companies take advantage of the environment. They show pictures of their products in the great outdoors, yet the things that they promote destroy the environment depicted in the ads. Thinks like detergents and soaps advertised to smell like "fresh water" or "rain" eventually make it into pristine rivers and waterways. The companies are working to destroy the very environment the companies use in their advertising. Rather than showing their products on the background of a beautiful natural backdrop they should show the problems their products cause.

And here is another student discussing Bill McKibben. McKibben's piece is more complex than the other readings in this lesson, and can be a bit challenging for students, but my student Michelle provides a strong analysis of his arguments:

Student Writing: Response Papers

by Michelle Ott

His concerns have now become my concerns. He feels as though society is drifting into a world of electronics as we drift away from the natural world. Although the television can often be extremely informative, there are lessons that only the natural world can teach us. As McKibben states, "There are lessons-small lessons, enormous lessons, lessons that may be crucial to the planet's persistence as a green and diverse place and also to the happiness of its inhabitants – that nature teaches and TV can't" (23). We have progressed into an age where the average human being obtains most of his knowledge about the world from the television. This concept makes me very concerned. Is it not true that the media could, if it so desired, brainwash us in such a way to believe what they want us to believe, to act the way they want us to act. Our actions are inspired by our thoughts and we form our thoughts from information we obtain. Through the language of our media, we can be influenced to act in whatever way the producers deem fit. We are slowly being disconnected from the natural world as we are influenced by the media.

When students come to class after reading and writing their response papers, I review some key concepts from the materials including examples of strategies used in media and in discourse to persuade or influence opinion (see the online resource for the slideshow I use).

Once we've reviewed these concepts, I share some examples of media products for students to analyze. First I show a series of video clips in which a particular interest group discusses an environmental issue. For each one, we watch the clip, then I ask students to analyze what they saw and what strategies were at work in the clip (see the online resource for links to these clips). We do this as a class, often replaying the clip two or three times. Students generally do very well at pointing out both visual and rhetorical strategies, noticing when one person is framed on the screen larger than another, when phrases are used that position the viewer to align with or against someone, when images flash in the background during a speech or narrative, when statistics are used deceptively, etc.

Next I show either news video clips, newspaper headlines, or news magazine covers and ask students which types of news bias might be influencing what is covered in these pieces. Following this I either share video clips of advertisements to analyze, or if students have laptops with them, I ask students to find advertisements they would like to analyze and email me the links. I pull up these videos on the classroom projector, and we talk as a class about what desires or needs the ads are attempting to appeal to (drawing from the list in the reading by Biagi, including such needs as "need for affiliation," "need for attention," and "need to feel safe"). After this full class discussion, you may have students break into groups and select additional media clips to analyze.

Overall, students should get a clear picture from this lesson that commercial media can have a strong influence in society, and that the contents of that media should be consumed critically, not unquestioningly. To demonstrate students developing these sorts of insights, I share part of a reply that an anonymous student wrote to a classmate's response paper:

> **Student Writing: Response Paper Reply**
>
> I think this also ties in with the idea that our language and discourse reflect society's norms. Since ads can strongly influence human action and ideas, if the ads were to remove their capitalist/profit-seeking motives, I think ad [*sic*] could be utilized to promote more effective sustainability. However, this is really unlikely and I think your concluding advice is correct: we should be more cautious of what is presented to us.

There are other activities I sometimes include in this lesson as well. For example, I have had students critique films on environmental issues, analyzing what messages about the natural world, and about environmental problems, are conveyed in the films and tying this in to their knowledge of media practices gained in this lesson. I've had students study the type and frequency of news stories about environmental issues in television or print news. And I have had students work with local environmental organizations to design media campaigns around an issue the organization is working on. For more information on these activities, see the online resource for this book.

Products and Assessment

The primary vehicles for evaluating students' achievement of the objectives for this lesson are their written response papers and their participation in class discussion. During class discussion they will analyze video clips – through this analysis they should demonstrate their grasp of the media strategies discussed in the readings and in class, and show critical thinking about media products.

I use the following grading criteria to evaluate student work for this lesson. I have, in different classes, formatted these criteria as a grading rubric or listed them in my course syllabus. How to format and apply the grading criteria will depend on individual context, school requirements, and teacher preference.

Grading Criteria

- Student demonstrates an understanding of rhetorical and visual strategies used in the media to frame information, persuade, sell, and secure or maintain agreement with particular ideologies.
- Student demonstrates critical thinking about media products and critical analysis of strategies used in the media.
- Student applies knowledge of forms of bias in news media, advertising strategies, and other media techniques to analyze media products.
- Student critically analyzes the motivations and interests behind media products.

- Student demonstrates critical analysis of specific words and phrases and their impact on our attitudes about other humans, nonhuman beings, and the natural world.
- Student demonstrates critical analysis of the arguments made by authors, how they play out in their lived experiences, and their relevance to social and environmental justice.
- Work demonstrates thorough reading and comprehension of assigned course texts.
- Work demonstrates personal reflection, critical thought, and insight into course texts.
- Arguments are clear, well-developed, and documented with evidence from texts.
- Work demonstrates critical analysis of course topics, questions, and subject-matter.
- Style, usage, format, grammar, imagery, and presentation support meaning and are intentional, creative, and original.
- Work meets all requirements of the assignment and is utilized to facilitate development of personal understanding.

EFFECT OF THE LESSONS: CRITICAL ANALYSIS OF DISCOURSE AND CULTURE

In any culture there will be dominant patterns of belief, dominant narratives that help us understand our world. These patterns will be embedded in language, discourse, and other symbolic resources. To make positive transformation in the world, we need to learn to recognize these patterns, be savvy about how they're communicated to us, and understand that they help structure what people assume to be "normal" or "natural" in ways that many individuals typically aren't fully aware of.

A major component of understanding these patterns of thought is to have the ability to critically analyze the influence of discourse on our thinking. Language and discourse operate as a powerful mechanism for structuring belief and action. The influence of discourse can reproduce harmful modes of perceiving and acting, or it can be applied to create positive transformation through reframing, relanguaging, and reconceptualizing ourselves and our interactions with the more-than-human world.

Related to gaining a critical understanding of the cultural impact of discourse, students should also gain wider knowledge of the underlying worldviews of their culture – the stories we tell ourselves about our purpose and our relationship with other beings. Environmental ethicist Mick Smith comments that our worldviews, ideologies, and discourses are not only conceptual "tools" but also "the framework[s] within which problems develop and proposed solutions are judged" (Smith 2001, 20). By understanding that more than one such framework exists, we can begin to recognize that the problems and proposed solutions that arise within our culture are not unalterable truths, but products of the belief systems that engendered them. In this way, expanded knowledge of the worldviews, foundational narratives, and practices of our culture and of a range of other cultures, past and present, provides us with a larger "toolbox" of cultural perspectives.

The lessons in this unit especially are intended to provide students with this larger toolbox; to help students become skilled investigators of language, discourse, media, and culture, who can identify discursive and ideological processes at work, analyze their impacts, and identify who benefits from or is harmed by them. (This critical analysis of culture and discourse should extend into all other lessons in the book, as well, of course.) As you teach the lessons in Unit 2, look for students to start recognizing that assumptions we may have come to think of as simply "given" are, in fact, culturally constructed, that they are not inherently true, and that they vary between cultures and over time. You can see that realization developing in this excerpt from a student discussing a reading from Lesson 2.3:

Student Writing: Response Papers

by Chelsie Bateman

Something else that Gee said that really pertains to this topic was, ". . . the meaning of words are composed of changing stories, knowledge, beliefs, and values that are encapsulated in cultural models, not definitions" (Gee, Pg 7). So, the labels we have given to women and the natural world do not define who or what they are. They are there because we put them there.

Also look for students to start pointing out the destructive impacts of certain beliefs, and look for them to begin the work of trying to conceive of alternatives – alternative language, alternative cultural metaphors, alternative beliefs. Here's an example of a student beginning to raise the question of better alternatives:

Student Writing: Response Papers

by Jacob Rosenborough

[O]ur culture is out of whack with its priorities and that a decent amount of our metaphors and sayings help reinforce this "out of whack-ness". My question is how do we fix this? What must we as a country and culture do to get our priorities on the right track, and what new metaphors and saying [*sic*] can we introduce and adopt to change the course we have gotten ourselves on? I don't think these questions have simple answers (if any at all) but it's clear that we hurt the environment with the map of metaphors we've created so something must be done.

Each lesson in this unit covers a different facet of our underlying cultural belief systems and the processes that shape them. Reading about different types of worldviews (as in Lesson 2.1) is useful to help students recognize that the dominant beliefs of their culture are not the only beliefs, or even the only commonly held beliefs throughout the world. Pairing this with reading a variety of creation myths from different continents gives students useful cultural perspective. Reading these myths is also an important chance for students to start recognizing that a culture's foundational belief systems can influence their attitudes and practices. Following this with an in-depth discussion of cultural metaphors in Lesson 2.2 gives students examples of common metaphors for the natural world, and the space to unpack each one and its potential consequences. Combining these with lessons 2.3–2.5 that highlight the role of language use and media in influencing people's beliefs about others, students can begin seeing examples of words and phrases that direct our attitudes toward historically oppressed human groups, nonhuman animals, and the land.

As students gain the tools to think critically about culture and discourse, they will also begin to feel more empowered to use the symbolic resources at their disposal in order to shape more healthy and sustainable patterns of thought and action:

Student Writing: Response Papers

by Michelle Ott

The power to change the world and better the environment is in our hands and in our words.

Arming students with these skills and engaging with them in this sort of conversation offers a powerful opportunity to institute a process of re-evaluation, through which we can begin to identify our motivations and beliefs and actively re-imagine new approaches grounded in care, justice, community, and interconnection. By forging a shared experience of dialogue and critical investigation into the cultural roots of our thinking about the natural world, students not only become more aware of the influences on their own and others' thinking, they also become empowered to actively reformulate modes of thinking in order to enact positive change.

UNIT 2 READINGS AND REFERENCES

Aristotle. (350 BCE) 2014. "History of Animals." Translated by D'Arcy Wentworth Thompson. *The Internet Classics Archive.* Book IX, Part 1, paragraphs 5–7.

Aristotle. (350 BCE) 2014. "Politics." Translated by Benjamin Jowett. *The Internet Classics Archive.* Part XII.

Berry, Wendell. 1995. *Another Turn of the Crank: Essays.* Washington, DC: Counterpoint.

Biagi, Shirley. 2011. *Media Impact: An Introduction to Mass Media.* 10th ed. Boston: Cengage Learning.

Black Elk. 2000. "The Hoop of the World." In *Environmental Discourse and Practice: A Reader,* edited by Lisa M. Benton and John Rennie Short, 257. Malden, MA: Wiley-Blackwell.

Capra, Fritjof. 2002. *The Hidden Connections.* London: HarperCollins.

Carson, Rachel. 2000. "The Obligation to Endure." In *Environmental Discourse and Practice: A Reader,* edited by Lisa M. Benton and John Rennie Short, 126–128. Malden, MA: Wiley-Blackwell.

Chief Seattle. [1855] 2000. "How Can One Sell the Air?: A Manifesto for the Earth." In *Environmental Discourse and Practice: A Reader,* edited by Lisa M. Benton and John Rennie Short, 12–13. Malden, MA: Wiley-Blackwell.

Cole, Thomas. 2000. "Essay on American Scenery (1835)." In *Environmental Discourse and Practice: A Reader,* edited by Lisa M. Benton and John Rennie Short, 87–90. Malden, MA: Wiley-Blackwell.

Corbett, Julia. 2002. "A Faint Green Sell: Advertising and the Natural World." In *Enviropop: Studies in Environmental Rhetoric and Popular Culture,* edited by Mark Meister and Phyllis M. Japp, 141–160. Westport, CT: Praeger.

Cox, Robert. 2010. *Environmental Communication and the Public Sphere.* 2nd ed. Thousand Oaks, CA: Sage Publications.

Dunayer, Joan. 2001. *Animal Equality: Language and Liberation.* Derwood, MD: Ryce.

Foreman, Dave. 2000. "Confessions of an Eco-Warrior." In *Environmental Discourse and Practice: A Reader,* edited by Lisa M. Benton and John Rennie Short. Malden, MA: Wiley-Blackwell.

Gee, James Paul. 1996. *Social Linguistics and Literacies: Ideology in Discourses.* 2nd ed. London: Taylor & Francis.

Glenn, Cathy B. 2004. "Constructing Consumables and Consent: A Critical Analysis of Factory Farm Industry Discourse." *Journal of Communication Inquiry* 28 (1): 63–81. doi:10.1177/0196859903258573.

Herrera, Juan Felipe. 2003. "Earth Chorus." In *From Totems to Hip-Hop: A Multicultural Anthology of Poetry Across the Americas, 1900–2002,* edited by Ishmael Reed, 30–32. New York: Thunder's Mouth Press.

Honeyborne, James. 2013. "Elephants Really Do Grieve like Us: They Shed Tears and Even Try to 'Bury' Their Dead – a Leading Wildlife Film-Maker Reveals How the Animals Are like Us." *Mail Online.* January 30. www.dailymail.co.uk/news/article-2270977/Elephants-really-grieve-like-They-shed-tears-try-bury-dead-leading-wildlife-film-maker-reveals-animals-like-us.html.

Hubbell, Sue. 1994. "Mites, Moths, Bats, and Mosquitoes." In *Reading the Environment,* edited by Melissa Walker, 1st ed., 161–164. New York: W. W. Norton.

Jefferson, Thomas. 2011. "Notes on the State of Virginia." *Electronic Text Center, University of Virginia Library.* February 21. http://web.archive.org/web/20110221131356/http://etext.lib.virginia.edu/etcbin/toccer-new2?id=JefVirg.sgm&images=images/modeng&data=/texts/english/modeng/parsed&tag=public&part=14&division=div1.

Klein, Naomi. 2010. "A Hole in the World." *The Nation.* June 24. www.thenation.com/article/36608/hole-world?page=0,0.

Lakoff, George, and Mark Johnson. 1980. *Metaphors We Live By.* Chicago: University of Chicago Press.

Leopold, Aldo. 1970. *A Sand County Almanac: With Essays on Conservation from Round River.* 1st Ballantine Books ed. New York: Ballantine Books.

Martusewicz, Rebecca A., Jeff Edmundson, and John Lupinacci. 2011. *EcoJustice Education: Toward Diverse, Democratic, and Sustainable Communities.* 1st ed. New York: Routledge.

McKibben, Bill. 1990. *The End of Nature.* 1st Anchor Books ed. New York: Anchor Books.

McKibben, Bill. 1993. *The Age of Missing Information.* New York: Plume.

Meadows, Donella. 1994. "What Is Biodiversity and Why Should We Care about It?" In *Reading the Environment,* edited by Melissa Walker, 1st ed., 149–151. New York: W. W. Norton.

Meadows, Donella. 1999. "Lines in the Mind." In *Our Land, Ourselves: Readings on People and Place,* edited by Peter Forbes, Ann Armbrecht, and Helen Whybrow, 2nd ed., 53–55. San Francisco: Trust for Public Land.

"Media/Political Bias." 2014. *Rhetorica.* Accessed October 28. http://rhetorica.net/bias.htm.

Merchant, Carolyn. 2000. "Ecofeminism." In *Environmental Discourse and Practice: A Reader,* edited by Lisa M. Benton and John Rennie Short, 209–213. Malden, MA: Wiley-Blackwell.

Mühlhäusler, Peter. 2003. *Language of Environment, Environment of Language: A Course in Ecolinguistics.* London: Battlebridge.

Muir, John. 2000. "A Voice for Wilderness (1901)." In *Environmental Discourse and Practice: A Reader,* edited by Lisa M. Benton and John Rennie Short, 102–104. Malden, MA: Wiley-Blackwell.

National Association Opposed to Woman Suffrage. 2014. "Vote NO on Woman Suffrage." Pamphlet reprinted in The Atlantic.com. Accessed October 11. www.theatlantic.com/sexes/archive/2012/11/vote-no-on-womens-suffrage-bizarre-reasons-for-not-letting-women-vote/264639/.

Nixon, Richard. 2000. "Message to Congress (1970)." In *Environmental Discourse and Practice: A Reader,* edited by Lisa M. Benton and John Rennie Short, 132–139. Malden, MA: Wiley-Blackwell.

Roosevelt, Theodore. 2000. "Conservation, Protection, Reclamation, and Irrigation (1901)." In *Environmental Discourse and Practice: A Reader,* edited by Lisa M. Benton and John Rennie Short. Malden, MA: Wiley-Blackwell.

Sanford, J. B. 2014. "Argument Against Women's Suffrage, 1911." San Francisco Public Library. Accessed October 11. http://sfpl.org/pdf/libraries/main/sfhistory/suffrageagainst.pdf.

Sire, James W. 1988. *The Universe Next Door: A Basic Worldview Catalog*. 2nd ed. Downers Grove, IL: InterVarsity Press.

Smith, Mick. 2001. *An Ethics of Place: Radical Ecology, Postmodernity, and Social Theory*. Albany, NY: State University of New York Press.

Snyder, S. 2014. "The Great Chain of Being." *Grand View University Faculty Pages*. Accessed October 28. http://faculty.grandview.edu/ssnyder/121/121%20great%20 chain.htm.

Soussan, Tania. 2004. "Scientist: Prairie Dogs Have Own Language." *redOrbit*. December 4. www.redorbit.com/news/display/?id=108412.

Sproul, Barbara C. 1991. *Primal Myths: Creation Myths Around the World*. 1st ed. HarperCollins ed. San Francisco: HarperSanFrancisco.

Swenson, May. 2003. "Weather." In *From Totems to Hip-Hop: A Multicultural Anthology of Poetry Across the Americas, 1900–2002*, edited by Ishmael Reed, 52–53. New York: Thunder's Mouth Press.

Szymborska, Wisława. 1998c. "Water." In *Poems: New and Collected 1957–1997*, 58–59. San Diego: Harcourt.

Viegas, Jennifer. 2005. "Chickens Worry About the Future." *ABC Science*. July 15. www. abc.net.au/science/articles/2005/07/15/1415178.htm.

Walker, Alice. 2014. "Am I Blue." *The Westcoast Post*. Accessed October 28. http:// westcoastword.wordpress.com/2013/06/01/am-i-blue-by-alice-walker/.

Warren, Karen. 2000. *Ecofeminist Philosophy: A Western Perspective on What It Is and Why It Matters*. Lanham, MD: Rowman & Littlefield.

White, Lynn. 1967. "The Historical Roots of Our Ecologic Crisis." *Science* 155 (3767): 1203–1207.

Wilson, E. O. 1994. "Storm over the Amazon." In *Reading the Environment*, edited by Melissa Walker, 1st ed., 151–160. New York: W. W. Norton.

Wright, Ronald. 2005. *A Short History of Progress*. New York: Da Capo Press.

Unit 3

PLACE

D. H. Lawrence writes, "In every living thing there is a desire for love, for the relationship of unison with the rest of things." . . . If I choose not to become attached to nouns – a person, place, or thing – then when I refuse an intimate's love or hoard my spirit, when a known landscape is bought, sold, and developed, chained or grazed to a stubble, or a hawk is shot and hung by its feet on a barbed-wire fence, my heart cannot be broken because I never risked giving it away. . . . Our lack of intimacy with each other is in direct proportion to our lack of intimacy with the land. We have taken our love inside and abandoned the wild. . . . The land is love. Love is what we fear. To disengage from the earth is our own oppression.

—*Terry Tempest Williams, "Winter Solstice*
at the Moab Slough" (63–65)
Lesson 3.1

This unit is designed to help students explore the significance of place in their lives and in the lives of others. Through the readings and activities in these lessons, students should reflect on their own memories and experiences of places that are important to them, think about how these places have influenced their lives, and think deeply about how "place" and "home" are experienced today for humans and for other species, especially as a result of the trajectory of human construction and development. Lesson 3.1 includes readings in which authors explore personal experiences of connection to place and memories of their childhood homes, as well as readings examining one of the most common types of living space for people in the U.S. – suburbia – and how and why suburban sprawl has developed the way it has. Readings also consider the flip side of human expansion by looking at "habitat," a term our society uses to talk about the homes of other species.

Readings consider the effect of habitat loss on other species, and the complexity and fragility of growing and living as part of an intact ecosystem. In Lessons 3.2 and 3.3, students engage in projects to further explore their own "places," learning about where their family has lived in past generations and telling stories of a place in the natural world that is special to them.

Some key questions that the lessons in this unit should raise for students:

- How do the places where we spend our time shape who we are as people?
- What places are special to you and why? What memories do you have of these places?
- Why are American homes, towns, and cities organized the way they are? How does this influence our lives?
- How does our culture encourage us to feel about the places we live and spend time?
- Do Americans typically value connection to place? Why do you think our culture does or does not feel such connection is important?
- What does "home" mean for other species? How are other beings' homes affected by human actions?

LESSON 3.1: SENSE OF PLACE

The Meaning and Experience of Connecting to Place

About this Lesson

In this lesson students read literature that explores personal connection to place. They reflect on the memories and experiences that individuals have about particular meaningful places, and explore what it means to have a "sense of place." They also learn about how our experience of place has changed in the U.S. over time, and the impact of the development of suburbs as a common mode of development. Students also consider the loss of place experienced by other species as their habitat is destroyed through human expansion.

Lesson Objectives

Students will:

- Reflect critically on literature and poetry.
- Apply insights from literature and poetry to their own experiences.
- Compare others' experiences of connection to place with their own.
- Understand the concept "sense of place" and reflect on its importance.
- Learn about urban and suburban development and analyze the benefits and differences of the two.
- Learn about the history of how suburbia developed in the U.S.
- Think critically about cultural attitudes toward place.
- Compose original analytical writing.

Lesson Activities at a Glance:

1. Students read literary passages, essays, and poems that explore experiences of place.
2. Students write response papers outside of class.
3. In-class discussion of insights from the readings and personal experiences with place.

Key Content Area Skills: reading and analyzing literature, essays, and poetry, acquiring historical knowledge, writing analytical and reflective essays.

Texts Used in this Lesson

Passages from *Reading the Environment*, edited by Melissa Walker:

"The World Is Places" by Gary Snyder, 88–91
"Land Where the Rivers Meet" by Annie Dillard, 92–93
"The Place Where I Was Born" by Alice Walker, 94–97
"The Lake Rock" by Ann Zwinger, 100–105

Aldo Leopold, *A Sand County Almanac*: "Good Oak," 6–19

George Ella Lyon, "Where I'm From," in *The United States of Poetry*, 22–23

Terry Tempest Williams, *An Unspoken Hunger*:

> "Winter Solstice at the Moab Slough," 61–65
> "Stone Creek Woman," 67–72
> "Yellowstone: The Erotics of Place," 81–87

Luci Tapahonso, *Blue Horses Rush In*: "A Song for the Direction of North," 5–6

Duany, Plater-Zyberk, and Speck, *Suburban Nation: The Rise of Sprawl and the Decline of the American Dream,* 3–20

James Howard Kunstler, *The Geography of Nowhere: The Rise and Decline of America's Man-Made Landscape*, 39–42, 85–94, 113–121

David Suzuki and Wayne Grady, *Tree: A Life Story*, 9–15, 43–52, 71–74, 133–140, 156–164

Roger Harrabin, "World Wildlife Populations Halved in 40 Years – Report," *BBC News*

Optional: Wendell Berry, *Another Turn of the Crank*, 46–55

Optional: David Abram, *The Spell of the Sensuous*, 137–179

Optional: William Cronon, "The Trouble with Wilderness; or, Getting Back to the Wrong Nature," in *Uncommon Ground*, 69–90

Optional: Leslie Marmon Silko, *Ceremony*, 1–45, 203–204, 244–247 (or whole novel)

Note: Publication information for these readings is listed at the end of the unit. Sample passages are included in the full List of Readings at the end of the book.

My Procedure

I start by having students read the assigned passages and write response papers outside of class. The readings for this lesson cover many facets of our relationship with place, and you may wish to split them up over multiple class sessions or weeks. Some readings are short essays, stories, and poems in which the authors explore personal memories about a place that is meaningful to them or expressions of emotional connection to place. Some authors, like Walker, Zwinger, and Lyon, share personal details such as remembering a specific tree they felt connected to when they were young, or recounting a series of images and memories from their childhood home and backyard. Some, like Snyder and Williams, talk about why their love of the land is meaningful to them and what it means to be connected to place. From here the readings move into different topics: the passages by Duany, Plater-Zyberk, and Speck and by Kunstler discuss how the suburbs came to exist in the U.S., and what it means to live in our car-centric, separately zoned, modern suburban towns. And pieces like those by Aldo Leopold and Suzuki and Grady highlight that the world is home to other beings, not just humans, and that these beings have long lives rooted in specific places and relationships as well.

As students read these pieces, they start to discuss the significance of place in their own lives and the significance of place for culture. Here's an example of a student exploring these ideas as she discusses the reading by Gary Snyder:

Student Writing: Response Papers

by Michelle Ott

Snyder expresses his idea that society experiences life in places. Every place means something different as various emotions and feelings are attached to the places where we travel. We remember our journeys. We remember the ventures we took through the woods. Although we may have been young and naïve children at the time, these journeys created who we are. The places of our childhood remain in our memories and consequently create the individuals who we are. Do you think you would be the same person you are today if you were raised somewhere else? Would you be a different person if you were in a different place and saw different things? Do you think those things would influence your morals, your perception, and your actions?

Students often respond to the personal memories in these passages by writing about memories of their own connections to individual trees, streams, and other beings and landscape elements. Then, as they read pieces like Aldo Leopold's "Good Oak" and Suzuki and Grady's *Tree: A Life Story*, they link their memories of these personal connections with larger consideration for the value and life experiences of other beings. Here a student discusses Leopold:

Student Writing: Response Papers

by Rebecca Postowski

It's fascinating to me thinking about how long trees are really around and what they have witnessed. I mean, think about it. Most of the time, at least once in your day, you are bound to come across a tree at some point, or even hundreds. I guess that sense of wonder astonishes me trying to put together how much does a tree know about the world. . . . I agree with Leopold's view that trees due [*sic*] in fact carry so many stories that occur in nature, and perhaps that makes them the wisest living organism on earth. . . . I really connected to this writing because I remember a tall oak tree at my elementary school where I always spent my recess time. The oak tree was where I first learned how to climb a tree, how I understood the life cycle of butterflies (tons of caterpillars scurried on the tree), and where I really bonded with nature. That tree held so many memories to me, and it really broke my heart when I found out the tree was cut down due the building of a new elementary school.

Reflecting on connection to other beings and places is a vital part of this lesson, but *lack of connection* is an equally important topic during this lesson. For most Americans today, the experience of place may not be primarily about bonds with trees or streams. In fact, the dominant experience may be more of "placeless-ness" – feeling uprooted, disconnected. Students discuss this disconnection quite a bit as they read the assigned texts for this unit. Exploring the phenomenon of the suburbs is an important way to help understand this experience of discon-nection from place; the readings by Duany, Plater-Zyberk, and Speck and by Kunstler argue that suburban sprawl is a major factor in disconnection from place. They suggest that suburbia, with its isolated zoning, car-centric structure, and homogeneity, does not take into account local ecosystems or traditions. Here's an example of an anonymous student reflecting on these problems of suburbia:

Student Writing: Response Papers

Kunstler attributes society's lack of understanding of landscape and where we live because people spend more time sitting in a car trying to get to a destina-tion rather that living in the space they are already in. People may argue that cars and other automobiles and machinery have improved outlets for com-munication and interaction but the reading holds the opposite position. Cars lead to less walking, which in turn means less interaction with nature and surroundings. The landscape now is man-made. Home lawns must be cut and maintained meticulously. . . . The roads have carved into the body of the earth and left it scared [sic] with its segregated areas for residents, consumption and recreation only connected by roads. With a landscape that is discontinuous and only accessible through certain roads, it is no wonder why people feel lost and uncertain of where they belong.

Students also note other ways that disconnection from place and from the land has become a common experience. Many express that they believe our culture is structured to discourage or prevent connection to place and to the natural world. Students are often very struck by one of Terry Tempest Williams' stories, "Winter Solstice at the Moab Slough," in which she discusses humankind's need for love and connection with the natural world, and how we have shut ourselves off from this need. Student Ashley Sweet comments, "The idea that separating ourselves from the land actually oppresses us was really powerful." Here's an excerpt in which another student writes about this piece:

Student Writing: Response Papers

by Will Fejes

On page 63, Williams quotes D.H. Lawrence,"In every living thing there is a desire for love, for a relationship of unison with the rest of things." I took this quote into deep consideration, it is one of the best quotes I have read in all of our readings so far. I think that is [sic] truly speaks to a specific attitude though. I say this because I do believe the quote is true, but I think that once again our

> modern day society has the ability to tear apart everything that we once stand
> [*sic*] for. What I am trying to say is that every living thing may have once had
> the desire for love and unison but something got in the way of those feelings
> and was ruled more important by the person and by society. In modern society
> we can be placed in situations in which we must choose environment over self
> gain or we are forced to choose self gain over environment.

Another important point for students to begin to notice is that human use
of space often conflicts with other species' use of space in the modern industrial world. As humans expand, build suburbs, delineate more and more space
as "ours," we take away the homes of other beings. Responding to William
Cronon's essay "The Trouble with Wilderness," one student writes on this topic:

Student Writing: Response Papers

by Chelsie Bateman

Humans made the wilderness so wild because of their want of escape, which
has protected the animals that live in these areas. But as the human population is growing and people do not want to live on farms or in the city, these
areas are becoming endangered. "Wilderness had once been the antithesis of
all that was orderly and good – it had been the darkness, one might say, on the
far side of the garden wall . . . " (Pg 5) It does not seem like the wilderness is
that dark, elusive place anymore. As people move in, the animals move out
(like the tigers, but instead of "moving out", they can't survive). So, it is like
humans are the darkness, taking over everything and not thinking about the
consequences and the wilderness is just another place.

Students begin to explore each of these topics in their response papers outside of class, and we continue to discuss them once students come to class. I ask
students about their reactions to the readings, and to share some of their own
memories of places that are meaningful to them. I ask if they can relate to the
experiences the authors describe.

One way to help get this discussion going in class is to discuss the poem "Where
I'm From" by George Ella Lyon. This piece is a great example to get students thinking
about their own memories – in it, the author describes a series of childhood objects,
foods, and family events. I have students analyze the piece and discuss it in class; you
also may consider having students write similar poems of their own. Linda Christensen
(2000) has written a terrific lesson around this poem in which students write their own
versions of it, modeling their poems after Lyons – see *Reading, Writing, and Rising Up:
Teaching about Social Justice and the Power of the Written Word* for her lesson.

I next ask students how important the places where they have lived and spent
time are to who they are as people. We talk about the connection between place
and identity and what it means to be "from" a particular place. We talk about

how often students have moved to new homes during their lives, and if they feel connected to the place they live now. We talk about their families' attitudes toward home and place, and about society's attitudes toward place. I ask if my students think that "sense of place" is something our society values today. I also ask how many students live or have lived in the suburbs and what mode of transportation they typically use to get where they need to go. We discuss the ramifications of living in a car-dependent society, and the difference between modes of development in which people must drive to get from home to work to grocery shopping, etc. versus more high-density modes of development in which people can walk and take public transit to most destinations.

Over the course of this lesson students should be demonstrating, both in their response papers and in class discussion, that they are thinking about how place has shaped their identity, about why and how they've experienced place the way they have, and about how our society has come to value, construct, and think about places – both the ones we live in and the ones other beings call home as well.

Products and Assessment

Students' written response papers and their participation in class discussion are the primary products for assessing their achievement of this lesson's objectives.

I use the following grading criteria to evaluate student work for this lesson. I have, in different classes, formatted these criteria as a grading rubric or listed them in my course syllabus. How to format and apply the grading criteria will depend on individual context, school requirements, and teacher preference.

Grading Criteria

- Student demonstrates critical analysis of literature and poetry.
- Student demonstrates thoughtful reflection about personal memories and experiences of place.
- Student demonstrates understanding of U.S. historical events surrounding the development of the automobile and suburbia.
- Student demonstrates critical analysis of the pros and cons of varying modes of development, including urban versus suburban structures.
- Student draws thoughtful connections between course readings and her/his own experience.
- Student draws connections between social issues around place and human uses of natural spaces.
- Student demonstrates awareness of the implications of habitat loss for other species.
- Student demonstrates critical analysis of the arguments made by authors, how they play out in their lived experiences, and their relevance to social and environmental justice.

- Work demonstrates thorough reading and comprehension of assigned course texts.
- Work demonstrates personal reflection, critical thought, and insight into course texts.
- Arguments are clear, well-developed, and documented with evidence from texts.
- Work demonstrates critical analysis of course topics, questions, and subject-matter.
- Style, usage, format, grammar, imagery, and presentation support meaning and are intentional, creative, and original.
- Work meets all requirements of the assignment and is utilized to facilitate development of personal understanding.

LESSON 3.2: FAMILY HISTORY OF PLACE

About this Lesson

In this lesson students follow up on their reading about sense of place by investigating their own families' histories of place. Each student will interview a family member and develop a list of places their family has lived in recent generations and family memories of those places.

Lesson Objectives

Students will:

- Use personal interviews to gain knowledge of family history.
- Reflect on their family history in relation to their own sense of place.
- Link their families' history of place to understandings about experiences of place in U.S. culture.

Lesson Activities at a Glance

1. Students interview a family member outside of class.
2. Students write a summary and reflection of this interview.
3. Optional: students share their interview summaries and reflections with the class.

Key Content Area Skills: conducting personal interviews, reflecting on ethnographic family history, writing summaries and personal reflection essays.

Texts Used in this Lesson

There are no new assigned texts for this lesson. Refer to the readings assigned in Lesson 3.1.

My Procedure

This lesson follows directly from Lesson 3.1. After students have discussed the readings assigned in that lesson in which authors reflect on meaningful places, and have begun thinking about special places from their own lives, I begin this lesson by asking students what they know about the places where their own families have lived. I ask them how many places they remember living themselves – many of my students have already moved multiples times in their lives. Then I ask them where their family members lived in past generations, as far as they know.

Next I introduce an interview assignment in which they'll speak to an older relative and find out about their family's "history of place":

Writing Assignment: Family History of Place

Conduct an informal interview with a relative of a different generation than your own – preferably a grandparent or other elder. Ask them about every location they know of where your family and your ancestors have lived. Have them be as specific as possible, with the names of the country, region, state, and town/city if they know them. Ask them how long your family members lived in each of these places, to the best of their knowledge. Ask them to describe details of the places they themselves have lived – what home or homes were most meaningful to them, what do they remember about what their homes and towns looked like, what activities did they participate in when they lived there, what was the landscape and weather like, what were their neighbors like? If they know any stories about the places where previous generations lived, have them share what they've heard about what those places were like. Ask them how they would describe where they are "from" and how they would describe where your family is "from." Ask them what it means to them to be from this place or these places.

Take notes, or – with permission – record the interview. Then write a report on the interview, describing and analyzing what was said, reacting to what you learned, and relating it to the readings on place you've done for class. Think about the common experiences of place described in those readings, as well as the historical insights into why Americans do or do not value place. Connect these insights to what you've learned from your interview. Reports should be approximately 1000 words.

I don't share any examples of student writing for this assignment here, as my students' work contains personal details about their families and homes. Students should learn something new from their interviews, and should demonstrate that they're reflecting on and integrating this new information. For this assignment, encourage students to talk to someone with as much knowledge on the subject as possible, someone from another generation; a sibling will not be a good choice, for example. The written reports students complete should highlight some specific memories shared during the interview, and students should show thoughtful reflection on how their family has been connected to (or disconnected from) place over the generations, and how that may have affected their own lives. Students should also connect this to concepts you've read about and discussed in class during Lesson 3.1.

Finally, if any students are comfortable doing so, I ask them to share some of the information from their interviews in class, and talk a bit about what they took away from the experience.

Products and Assessment

The product students create for this lesson is the interview write-ups they submit after completing their interviews. These should show critical thinking about the importance of place and how their own family fits into the patterns of connection and disconnection from place you've discussed in Lesson 3.1.

I use the following grading criteria to evaluate student work for this lesson. I have, in different classes, formatted these criteria as a grading rubric or listed them in my course syllabus. How to format and apply the grading criteria will depend on individual context, school requirements, and teacher preference.

Grading Criteria

- Student conducts a thoughtful interview and relates the experiences of a family member to insights from course readings.
- Student demonstrates thoughtful reflection about personal memories and experiences of place and cultural attitudes toward the value of place.
- Student draws thoughtful connections between course readings and her/his own experience.
- Student demonstrates critical analysis of the arguments made by authors, how they play out in their lived experiences, and their relevance to social and environmental justice.
- Work demonstrates thorough reading and comprehension of assigned course texts.
- Work demonstrates personal reflection, critical thought, and insight into course texts.
- Arguments are clear, well-developed, and documented with evidence from texts.
- Work demonstrates critical analysis of course topics, questions, and subject-matter.
- Style, usage, format, grammar, imagery, and presentation support meaning and are intentional, creative, and original.
- Work meets all requirements of the assignment and is utilized to facilitate development of personal understanding.

LESSON 3.3: DIGITAL STORY OF PLACE

About this Lesson

In this lesson students employ the insights they have gained from Lessons 3.1 and 3.2, conduct research, and create a "digital story" in which they profile a particular place in the natural world that has special meaning for them.

Lesson Objectives

Students will:

- Reflect on personal experiences with place.
- Research the history and ecology of a specific place.
- Plan, create, and present a digital story, containing photographs, music, and a narrative, reflecting on the significance of a specific place for themselves and others.
- Employ empathy to imagine the value of a place to other people and other beings.

Lesson Activities at a Glance

1. Students research the history and ecology of a particular place in the natural world.
2. Students create a digital story profiling this place and its significance.
3. Students present digital stories in class.

Key Content Area Skills: researching history and ecology, writing reflective narratives, creating multi-media presentations, using digital media to reflect on and communicate personal experiences.

Texts Used in this Lesson

There are no new assigned texts for this lesson; refer to the readings in Lesson 3.1.

My Procedure

This lesson requires students to select a place in the natural world that has particular significance to them. I first introduce this assignment well in advance of its due date, giving students notice to be thinking about a place in the natural world they feel connected to, a place that they like to visit or have special memories of. I tell them to either track down photographs they and their family have taken of this place in the past, or to arrange to visit the place at least once or several times during the semester in order to observe and take pictures (or both). Later in the semester, once we have begun Unit 3, I give them the full assignment. It can be found in the box below.

Digital Story Assignment

Create a digital story describing a place that you feel connected to. You may select a place from your home area or a place you knew as a child, or you may choose a spot you visit now, such as a spot on or near campus. Keep in mind that if you select a place that isn't local, you'll need to be able to remember this place well enough to write about it from a distance, and you'll need access to photographs you have taken of the place. If you choose a local spot, spend some time each week sitting in the location you've chosen and get to know it. Watch how it looks at different times of day and as the season begins to change, watch the other people and creatures that occupy the space, notice how it interacts with your senses (what sounds, colors, smells, temperatures, textures, etc. do you notice when you're there?), and think about how you feel when you're in this place. Take notes each time you visit the spot. Also research the place you've chosen and find out some details about its history. What did it look like 200 years ago? How about 100? 50? 10? Look into the ecology of the place as well, and any modern threats that may affect this place in the future.

Then develop a digital story that introduces the class to this place. Describe the features of this place, both physical and emotional. What is the landscape? What is the local ecosystem? What plants grow there, and what nonhuman animals live there? What watershed is this place a part of? How have humans interacted with and changed this place over time? What have humans built here? Tell us the biography of this place, its history, but also tell us about its spirit. Think of this place as a "repository of meaning," a source of memories, experiences, and various types of resources for you and others. Think about what those resources are: What have you gained from this place? (Shelter? Space for play? An opportunity to learn, to grow, to spend time with loved ones? Beauty? Relaxation? A link to the rhythms of nature?) And what have others (both human and nonhuman) gained from this place? Think about who values and has valued this place, and why. What have you and others brought to this place? How have you interacted with the place, and what does it mean to you? Do your own "ecological identity work" and explore your relationship with this place.

Your digital story should be 5–7 minutes in length and should include both pictures and narration. If you can, locate historical photos as well as photos of the place as it looks today. You may also want to take your own photographs that express the character of this place and what it's like from your perspective. The narration can either take the form of text you speak live that accompanies a slide show, or of a recorded voiceover. You may want to narrate your story from your own perspective, or you may consider narrating your story from the perspective of someone else who has experienced this place: a plant or nonhuman animal living there, a building that has been built there, or even from the perspective of the place itself.

For this assignment, I ask students to focus on a place in the "natural world." This request has a good deal of gray area. Exploring our personal connections to city neighborhoods, apartment buildings, and other human-built structures is worthwhile, but for this digital story I want to make sure students take the time to reflect on how they feel when they are in a space that is not part of the built environment, when they are surrounded by other living beings beyond humans. I want them to think about how other beings use and think about this space, how they share it with one another and with humans, and how humans have influenced this space over time. For this reason, I ask that students select a place that is not primarily of human origin.

If students don't already have some place in the natural world that they feel connected to, I ask them to become familiar with a place over the early weeks of the semester prior to this assignment. It doesn't have to be a grand place, it can be a spot under a tree on the school grounds or at a local park. These places may not be part of "wild" nature, but there are other beings living and using them beyond humans. I just want students to have the experience of spending time and thinking about a space outside of the entirely human-built environment.

After I introduce this assignment, I usually show an example of a digital story created by a student from a previous semester (with their permission) in order to give my students a sense of what is possible. I then periodically check in with students about how their work on the assignment is progressing; we talk about what research they need to do and how to cite it, how to cite any photographs from the public domain that they use, and how to include sensory details to convey the feeling of being in their place.

I allow students to present their digital stories either as slideshows using software such as PowerPoint, or as movie files using software like iMovie or Windows Movie Maker. If they create a slideshow, they read their "narrative," the script or story they have written that accompanies their visuals, live in class as the slideshow plays. If they use movie software, they pre-record themselves reading their script as part of the soundtrack to the movie, playing it all as one video file. My students are typically more well-versed and experienced at making short movies than I am, and they rarely need any technical help. However, for links to tutorials and further information about digital stories should you or your students need assistance, see the online resource for this book.

In the box below I've excerpted part of the narrative from an anonymous student's digital story. This is the voiceover component of her digital story, minus the visual presentation of course. I've removed any personally identifying details, but tried to include enough of her narrative to demonstrate the elements of the assignment.

Student Writing: Digital Story Narrative

This place was . . . a home and sanctuary with whom [*sic*] I had built a strong relationship . . . The land is part of two watersheds, the little and middle Patuxents . . . which connect to the Chesapeake Bay . . . My special place is . . . named after a historic mill. The barns of the . . . dairy farm are still standing in the village center today. . . . As a child, the yard and neighborhood was a gigantic playground, where I found the most interesting things to do and learn. Unlike the housing developments today that wipe out all the trees, at [my neighborhood], as long as I can remember, the trees, nature, and sky toppled [*sic*] over the community. In the front of the house was a significant cherry blossom tree. I would lie underneath its shade gazing into the branches. Depending on the season, I'd see the blooms and blossoms of velvety baby pink petals, or the luscious waxy deep green leaves, or the ever-changing hues of the red, yellow, and orange. The sunlight would tinker through making a breath-catching kaleidoscope that would lull me into a trance. . . . Some days the family of birds nested in the middle branches would peak their heads out. . . . Their chirping added to the musical melody of the wind brushing up against the leaves. . . . In the fall, I would love playing in the fallen leaves of the maples and oaks that lined the sidewalk. . . . The piles, taller than me, would crunch and rustle under my feet. My whole body would be submerged and I would come back up for air with bits of dry leaves tangled in my hair. The air was crisp and the sunlight bright, casting a golden glow around me. . . . Winter is my favorite season, and that's mainly due to the precious memories I have at [my childhood house]. Starting from a younger age I was exposed to this magical white substance. Each year I'd taste a sample of the first snowfall, following an old superstition which, according to my grandmother, meant I would receive good health and luck. . . . Like my mother did as a child, I built countless numbers of igloos, snowballs, snowmen, and snow angels. . . . Out into the snow, my footprints would usually be next to the fresh ones of the fox that lived a few trees down, telling me that I wasn't the only one who cherished the snow's cooling touch. . . . Here I climbed the trees all the way up into the sky, with the pinecones falling down on me and the sap sticking to my hands. Here I would venture into the small pond and open space in the back yard. . . . The space seemed endless in my eyes. . . . Until recently, I did not know that the space was used for a petroleum pipeline. I thought no human could corrupt my special place. . . . The cicadas and spiders made their comfy and intricately webbed homes here. The ants, grasshoppers, and fireflies would never be too far away either. I would help my grandparents plant and tend their gardens around the house. The bright tulips would open up in the spring, attracting buzzing bees and beautiful butterflies. . . . The gardens had fresh lettuce, tomatoes, chives, squashes, sesame leaves, and melons that would be on my dinner plate, if a rabbit hadn't gotten to it first. From this experience I learned how giving and providing the earth is. Gardening taught me the lessons of responsibility, patience, and the tenacity needed to live a life in harmony with nature. . . . I can't help but think it is lonely without me. Although I am physically distant, I feel as if a part of me is still living here, lingering in that land.

When students have completed their digital stories, they present them in class. After each presentation, we take a minute to talk about the place the student showed us in their digital story and I ask if any classmates have questions or comments. At the end of these presentations, we also engage in a short wrap-up conversation, asking: What did students notice about these digital stories as a whole? Were there certain things they related to about other people's digital stories? Does this tell us anything about place and what it means to us?

Note: at the end of Unit 3, I often include another lesson in which I ask students to learn about a local environmental organization working on an issue in our bioregion. I have students interview someone from this organization and give presentations in class. More on this and other extra assignments can be found in the online resource for this book.

Products and Assessment

The primary product for this lesson is the digital story that students present in class. These digital stories should show critical thinking about the importance of place, should involve creative use of language to express their personal feelings and share what it feels like to be in the place, and should demonstrate historical and ecological research into their chosen place.

I use the following grading criteria to evaluate student work for this lesson. I have, in different classes, formatted these criteria as a grading rubric or listed them in my course syllabus. How to format and apply the grading criteria will depend on individual context, school requirements, and teacher preference.

Grading Criteria

- Student demonstrates thoughtful reflection about personal memories and experiences of place.
- Student demonstrates research into the history and ecology of a particular place, and cites all sources.
- Student draws connections between social issues around place, human uses of natural spaces, and personal experiences.
- Student demonstrates awareness of the shared use of space by other species.
- Student demonstrates thoughtful reflection into the role of place in personal identity.
- Student demonstrates empathy in imagining the significance of a particular place to other beings.
- Work demonstrates thorough reading and comprehension of assigned course texts.

- Work demonstrates personal reflection, critical thought, and insight into course texts.
- Arguments are clear, well-developed, and documented with evidence from texts.
- Work demonstrates critical analysis of course topics, questions, and subject-matter.
- Style, usage, format, grammar, imagery, and presentation support meaning and are intentional, creative, and original.
- Work meets all requirements of the assignment and is utilized to facilitate development of personal understanding.

EFFECT OF THE LESSONS: ECOLOGICAL IDENTITY AND CONNECTION TO PLACE

To be connected to place is to feel that we belong to something larger than ourselves, and to link our identity and our welfare to that of others. When we feel that we come from and are a part of the land, when we participate with the living earth around us, we view our own interests as inextricable components of a larger whole. Educational theorist Etienne Wenger (1999) suggests that through experiences of kinship and mutual recognition we find meaning and generate a sense of belonging and of self. He says, "In this experience of mutuality, participation is a source of identity. By recognizing the mutuality of our participation, we become part of each other" (56). Exploring this participation and reawakening ourselves to our belonging to place may be essential in helping us to reformulate our approaches to the world.

We must depend on one another and on the land for survival. Being part of a community, with each other and with the natural world, is key to our well-being as individuals and as a species. And it is equally vital that we *feel* and acknowledge these bonds that shape us; as David Orr (1992) argues, "even a thorough knowledge of the facts of life and of the threats to it will not save us in the absence of the feeling of kinship with life of the sort that cannot entirely be put into words" (87). Orr and others suggest that to truly care we must also form a bond, an emotional and intuitive connection, to others and to the living world. Forming such bonds internalizes and naturalizes the knowledge that our welfare is linked to the welfare of others. It also allows us to explore the full range of ways in which we are linked to other beings.

Recognizing our bonds with the earth and feeling ourselves to be part of a community with others and with the surrounding land can be facilitated by engaging in what Mitchell Thomashow (1995) has called "ecological identity" work. Doing this work means exploring the history of a place, one's own history in relation to that place, and one's personal experiences in that place. This exploration is one goal of Unit 3. The lessons in this unit should help students reflect on where they come from, and recognize that their heritage is not only about family and community, but also about their connections to the natural world. Paul Lindholdt (1999) says of this, "The analysis of oneself as a product of place becomes a challenge to be met by studying the particulars of one's ecological origins" (6).

In response to George Ella Lyon's poem assigned in Lesson 3.1, one anonymous student comments on this precise idea:

Student Writing: Response Papers

He is not focusing on himself and his own importance, but rather the place and the impact it has. I feel that we can all benefit from this type of thinking. We should not just look at a place as "ours" but rather as ourselves as a product of the place.

As students engage with the lessons in Unit 3, they start to recognize that they are connected to specific places in the world, and that these places have helped shape who they are. Reflecting on the emotional impact of places in their lives, and the role of those places in forming their personality and "self," students sometimes start to notice that they have had experiences of connection to place that they didn't realize:

Student Writing: Response Papers

by Ashley Sweet

At first I wasn't sure I understood how losing a place, a neighborhood, could damage a culture. The more I thought about it, the more I realized that I DID understand. Places matter to the people who interact with them. Land itself has always impacted me personally, but until this course I was highly unaware of how the places I've been have helped shape who I am. As a child (frankly, even as an adult) every time I move homes I feel uncomfortable, shaken and unsettled for a while. When I was little it was easy to just cry over leaving my room, my yard, and my neighborhood friends. But now as an adult I can relate to what my home, not just the house but the neighborhood, the city around it, means to me. It's familiar, it's what I associate with comfort and acceptance and belonging. It's the only place where being me is all I have to do. I know my neighbors and they know me. I know the flowers that grow in the yard, the way the wind blows my neighbor's leaves into my yard. I'm familiar with the road I live on, the place where the cement isn't even. My kids remember the patterns in the sidewalk, where the cracks are, where the caterpillars live, where you have to watch your step. It's easy to feel like it's just a place, like any other place. But it isn't. A neighborhood, a community means something.

As students think and learn about their own "ecological identity" throughout this unit, they should also be thinking about the social forces that influence our modern experience of place. There are reasons that many of my students hadn't thought about places that were meaningful to them, or why place matters in people's lives, prior to the assignments in these lessons. Consumer-oriented values that do not ascribe meaning to our surroundings beyond their role as status symbol and possession, car-centric lifestyles, and suburban development have become hallmarks of the American experience of place – or lack of experience of place. It's important to interrogate how this shapes our sense of identity, connection, and community. Students often reflect deeply on *disconnection* during this unit, as they consider the ways that modern lifestyles have isolated us from feeling connected to the land or having a sense of belonging to a particular place.

One of my students, Michelle Ott, comments, "We are destroying our planet; however, we are also destroying our connection to the world." Here's a key insight, that isolating ourselves from our connection to place is harmful both to us and to the natural world.

The harm we are doing to the natural world, and the rights of other beings to maintain their own homes, are also important topics for students to explore in this unit. As humans spread out, other beings lose their homes to serve human interests. Students may begin to see the hypocrisy in ignoring the well-being of other creatures as we pursue our own "space." Students may also start to notice that our culture often designates particular areas as "wild," isolating ourselves from those places while at the same time assigning the rest of the land outside these "wild" places to human ownership and control. Here's an example where a student raises some very interesting insights on this point:

Student Writing: Response Papers

by Heather Harshbarger

The main thing that humans need to understand is that nonhumans, especially the rooted population, have no sense of boundaries and do not follow those that humans have set. In the The World is Places reading by Gary Snyder, it states "Every region has its wilderness," then follows "I have a friend who feels sometimes that the world is hostile to human life—he says it chills us and kills us," (p91). Every region has a trace part of nature to it, even if it lies solely in a domesticated animal such as a dog. Looking further into it, his friend saying this about the world contributes to the feeling of "wilderness" humans often refer to when they mention the wild. With this image of wilderness, we are able to distance ourselves from it and to think of it in a way that almost doesn't affect us. It's easier to cut down trees, throw waste away, and continue our lifestyle when we don't think it's affecting us in any direct way.

Thinking about home, community, and self as grounded in relationships with the natural world, students can learn about themselves and about modern U.S. society, and may come to form new perspectives on their connections to other animals and "the rooted population," as my student so beautifully puts it.

Students may also come to see that place and community can be important parts of the *solution* to many social and environmental problems. Feeling a close connection to one's human and nonhuman neighbors and being an active participant in community with these others can counteract the alienation that many of my students highlight as so prevalent in modern life, and it can empower people to work together to address issues through community-based action. Rather than believing, as modern mainstream society might have us believe, that we are all individual actors who can only solve problems through personal consumer choice, generating a strong connection to place provides opportunities to see ourselves as a larger force, revealing to us both the broader needs of the community and its broader power. For, as Fritjof Capra (2005) tells us, "Sustainability always involves a whole community. This is the profound lesson we need to learn from nature" (24).

UNIT 3 READINGS AND REFERENCES

Abram, David. 1997. *The Spell of the Sensuous: Perception and Language in a More Than-Human World*. 1st Vintage Books ed. New York: Vintage.

Berry, Wendell. 1995. *Another Turn of the Crank: Essays*. Washington, DC: Counterpoint.

Capra, Fritjof. 2005. "Speaking Nature's Language: Principles for Sustainability." In *Ecological Literacy: Educating Our Children for a Sustainable World*, edited by Michael K. Stone and Zenobia Barlow, 1st ed., 18–29. San Francisco: Sierra Club Books.

Christensen, Linda. 2000. *Reading, Writing, and Rising Up: Teaching About Social Justice and the Power of the Written Word*. Milwaukee: Rethinking Schools Ltd.

Cronon, William. 1995. "The Trouble with Wilderness; Or, Getting Back to the Wrong Nature." In *Uncommon Ground: Toward Reinventing Nature*, edited by William Cronon, 1st ed., 69–90. New York: W. W. Norton.

Dillard, Annie. 1994. "Land Where the Rivers Meet." In *Reading the Environment*, edited by Melissa Walker, 1st ed., 92–93. New York: W. W. Norton.

Duany, Andres, Elizabeth Plater-Zyberk, and Jeff Speck. 2000. *Suburban Nation: The Rise of Sprawl and the Decline of the American Dream*. New York: North Point Press.

Harrabin, Roger. 2014. "World Wildlife Populations 'Plummet.'" *BBC News*. September 30. www.bbc.com/news/science-environment-29418983.

Kunstler, James Howard. 1994. *The Geography of Nowhere: The Rise and Decline of America's Man-Made Landscape*. 1st Touchstone ed. New York: Simon & Schuster.

Leopold, Aldo. 1970. *A Sand County Almanac: With Essays on Conservation from Round River*. 1st Ballantine Books ed. New York: Ballantine Books.

Lindholdt, Paul. 1999. "Writing from a Sense of Place." *Journal of Environmental Education* 30 (4): 4–10.

Lyon, George Ella. 1996. "Where I'm From." In *United States of Poetry*, edited by Joshua Blum, Bob Holman, and Mark Pellington, 22–23. New York: Harry N. Abrams.

Orr, David W. 1992. *Ecological Literacy: Education and the Transition to a Postmodern World*. New York: State University of New York Press.

Silko, Leslie Marmon. 1986. *Ceremony*. New York: Penguin Books.

Snyder, Gary. 1994b. "The World Is Places." In *Reading the Environment*, edited by Melissa Walker, 1st ed., 88–91. New York: W. W. Norton.

Suzuki, David T., and Wayne Grady. 2004. *Tree: A Life Story*. Vancouver: Greystone Books.

Tapahonso, Luci. 1997. *Blue Horses Rush In: Poems and Stories*. Tucson: University of Arizona Press.

Thomashow, Mitchell. 1995. *Ecological Identity: Becoming a Reflective Environmentalist*. Cambridge, MA: MIT Press.

Walker, Alice. 1994. "The Place Where I Was Born." In *Reading the Environment*, edited by Melissa Walker, 1st ed., 94–98. New York: W. W. Norton.

Wenger, Etienne. 1999. *Communities of Practice: Learning, Meaning, and Identity*. 1st ed. New York: Cambridge University Press.

Williams, Terry Tempest. 1994. *An Unspoken Hunger: Stories from the Field*. New York: Pantheon Books.

Zwinger, Ann. 1994. "The Lake Rock." In *Reading the Environment*, edited by Melissa Walker, 1st ed., 100–105. New York: W. W. Norton.

Unit 4

FOOD

Most urban shoppers would tell you that food is produced on farms. But most of them do not know what farms, or what kinds of farms, or where the farms are. . . .

If one gained one's whole knowledge of food from . . . advertisements (as some presumably do), one would not know that the various edibles were ever living creatures, or that they all come from the soil, or that they were produced by work. The passive American consumer, sitting down to a meal of pre-prepared or fast food, confronts a platter covered with inert, anonymous substances that have been processed, dyed, breaded, sauced . . . and sanitized beyond resemblance to any part of any creature that ever lived. The products of nature and agriculture have been made, to all appearances, the products of industry. Both eater and eaten are thus in exile from biological reality. And the result is a kind of solitude, unprecedented in human experience, in which the eater may think of eating as, first, a purely commercial transaction between him and a supplier and then as a purely appetitive transaction between him and his food. . . .

Eaters, that is, must understand that eating takes place inescapably in the world, that it is inescapably an agricultural act, and how we eat determines, to a considerable extent, how the world is used.

—Wendell Berry, "The Pleasures of Eating"
Lesson 4.3

The lessons in this unit focus on our modern food system and its social and environmental impacts. This includes considering how food is grown, what effect modern industrial agriculture has on the environment, how those who grow, ship, and prepare food are treated, how modern processed food affects health, who does and does not have easy access to healthy food, and how other species are treated, including how nonhuman animals are treated in factory farms.

As they learn about these elements of the food they eat, students can start to recognize unjust global processes at work, and at the same time they can start to connect these global dynamics to deeply personal practices and phenomena, such as the choice of what they put into their own bodies, and the process of growing and harvesting from the land.

Some key questions that the lessons in this unit should raise for students:

- Where and how is the food you eat produced, and who is affected by its production?
- How does culture influence what choices we make about what to eat?
- How does gender, race, and income level influence what we choose to eat and what food we have access to?
- How do food practices in the U.S. affect other countries?
- How do national and international policies influence what foods are produced and how they are grown/made? How do business interests influence what foods are produced and how they are grown/made?
- How are other beings treated in the process of making our food, and is this treatment ethical?
- What impact does modern food production have on the natural world?
- What are the most healthy, ethical, and sustainable ways to grow and consume food?

LESSON 4.1: THE MODERN FOOD SYSTEM

How We Produce, Sell, and Think about Food in the U.S.

About this Lesson

In this lesson students begin to explore the food system in the U.S., including modern industrial agriculture, how foods are marketed, how culture, gender, and other social factors influence food practices, and how our food system affects the natural environment. They then reflect on this new knowledge using creative writing.

Lesson Objectives

Students will:

- Gain knowledge of modern food production practices.
- Gain knowledge of the role of culture in influencing what people eat.
- Gain knowledge of the impact of food production and consumption on other people, other animals, and the land.
- Use creative writing for personal and cultural reflection.

Lesson Activities at a Glance

1. Students read about aspects of the modern food system and its impact.
2. Students write response papers outside of class.
3. Students discuss and reflect on their own food attitudes and experiences.
4. Students use creative writing to explore their understanding of the role of food in social and environmental problems.

Key Content Area Skills: reading and analyzing nonfiction prose, writing analytical and reflective essays, writing poetry and creative prose.

Texts Used in this Lesson

Anna Lappé and Bryant Terry, *Grub*, 3–51

Katherine Parkin, "Campbell's Soup and the Long Shelf Life of Traditional Gender Roles," in *Kitchen Culture in America*, 51–64

B. W. Higman, *How Food Made History*, 143–158

John Ryan and Alan Thein Durning, *Stuff: The Secret Lives of Everyday Things*, 7–12

Wenonah Hauter, *Foodopoly: The Battle Over the Future of Food and Farming in America*, 39–61

Michael Pollan, *The Omnivore's Dilemma*, 15–31

Optional: Harvey Levenstein, *Paradox of Plenty*, 227–236

Optional: Raj Patel, *Stuffed and Starved*, 75–117

Optional: Waverly Root and Richard De Rochemont, *Eating in America*, 13–28, 42–67, 74–88

Optional: Fabio Parasecoli, *Bite Me: Food in Popular Culture*, 1–14, 85–102, 103–125

Note: Publication information for these readings is listed at the end of the unit. Sample passages are included in the full List of Readings at the end of the book.

My Procedure

Students begin by reading the assigned texts and writing response papers outside of class. These texts introduce students to some facets of modern industrial agriculture, pesticide use, factory farming, what modern processed foods are made of and how they are made, the role of big business in producing and selling food, and the ways that culture can shape attitudes toward what is appropriate to eat. If students aren't familiar with this information, they may find some of it shocking. Here's an example of one student's response to *Grub* by Lappé and Terry:

Student Writing: Response Papers

by Mallory Brooks

Did anyone else catch that there were traces of 200 CHEMICALS INSIDE THE BLOOD OF AN UMBILICAL CORD?! So let me get this straight. We have found that these chemicals can cause cancer and other serious health issues, we have seen traces of it in the blood that is going into a child that hasn't even had a chance to live and make choices of what he or she may want to eat, and this is still legal?! We are continuing this foolishness of blanketing things we plan on ingesting with poison. Literally poison. I mean, are you going to eat or drink the crap that they are spraying by itself? Um no, why? Oh because it's poison. Seems legit. All this in the name of progress and money. It is quite interesting isn't it.

Learning about the food they eat can be personal and unsettling, but also a subject students quickly become passionate about. It's also important that students recognize that their food choices, while personal, are at the same time heavily influenced by cultural forces. Values, beliefs, messages from advertising, the influence of moneyed interests, systemic factors that affect what foods are available, and embedded cultural notions of what is signaled by eating certain foods all play a role in shaping our food choices. Students should begin to understand each of these factors, from the economic to the cultural. Here's a sample of a student commenting on the economic forces at work and their negative consequences:

Student Writing: Response Papers

by Danny Clemens

[O]ur present system of agribusiness is continuously and flagrantly putting vulnerable, at-risk groups of individuals in danger.

And here's a sample of writing in which a student comments on the cultural influences affecting our food choices:

Student Writing: Response Papers

by Yiannis Balanos

As the "Bite Me" reading points out, "Food influences our lives as a relevant marker of power, cultural capital, class, gender, ethnic, and religious identities." (2). Since food is so pervasive to defining cultural lifestyles and so essential to our survival, it is expected that it would be so ingrained into our culture and gender roles.

When students come to class after completing these readings, I have them talk in small groups first, sharing their reactions to the readings and any details that surprised them. I also ask them to discuss any connections they made between the readings and their own lives, discuss thoughts on any of their own food attitudes or habits, and identify cultural influences on their habits. Students will have already started to voice these connections in their response papers, and small group discussion gives them a chance to share and expand on what they wrote in close conversation.

Once students have discussed in groups, I ask them to share some of what they talked about with the whole class, recapping and sharing a few interesting points from their discussions. I find that many of the groups will have raised the same points about details that surprised them from the readings. The amount of pesticides used to grow industrial produce, the small number of major corporations that own much of the food supply, the money poured into marketing processed foods – these are a few subjects that students have strong reactions to. Another topic of strong interest is the ingredients found in processed foods; for example, the number of corn derivatives used in everything from sodas to hamburgers. As student Mallory Brooks says:

Student Writing: Response Papers

by Mallory Brooks

We, as in Americans, are people of the corn. I don't mean like that movie (Children of the Corn), I mean WE ARE WHAT WE EAT AND ALL WE EAT IS CORN. It's in everything.

Once we've discussed students' reactions and connected them to their own habits and experiences, I introduce a creative writing assignment. This assignment is intended to give students a flexible platform to voice some of the issues they find most disturbing or vexing about our modern food system, and to further link what they're learning to their own lives. Students write poems, essays, or sometimes mix text and artwork for this assignment – I keep it very open, but you could be more specific with the requirements if you prefer. This assignment could be given to students at the end of this lesson, or at the end of Lesson 4.2 or 4.3; I often introduce this assignment during Lesson 4.1 and then let students work on it through 4.2, turning it in during or at the end of Lesson 4.3.

Writing Assignment: Reflective Writing on Food

Partway through this unit you will submit a piece of reflective writing exploring your reactions to the materials we've discussed in class and your own experiences with and attitudes toward food and food culture. This piece should be creative, and can take whatever form you select, including a first-person short story, essay, or poem. However, it must include a written component and should make use of insights from at least three class texts. Think about what insights you found most surprising or important from our readings so far. Also think about your own food choices, why you buy and eat the food you do, and any ways that you've come to think differently about your food habits after our class readings and discussions. Include a list of works cited at the end identifying which texts you referenced in your piece.

The following box contains an example of a poem written by a student for this assignment – or as my student calls it, here's his "wicked rap piece."

Student Writing: Reflective Writing on Food

by Dwarka Pazavelil

If we are what we chew
I'd go so far as to prove
That we're more than the food
That our farmers produce.
And the fruits of their labor
Don't exactly favor you.
So what is it that they do?
Let's savor that thought
And for a moment let's talk
About the thrown in chalk
And other additives added in,
Some grown in a lab in which
Managin' the results
and its damagin' effects
Affect the price we pay
In more than one way.
Sure, our bread is whiter
And seemingly lighter
But, higher is the fact
That it's a nutritional liar
With a side of disease
Added to that.
It is no Wonder
this style of bread
Has lost its thunder
. . . But not quite dead
As we are proficiently fed
Images and messages,
For most of our grub,
Incredibly misled
To believe
That that tub of butter
Won't put us on meds.

It's sometimes expressed
"An apple a day . . . "
Well, you know the rest
But, at the same time

It could flatline
Your troughs and crests.

Even academics who reject
To connect the body and mind
Keep their body in mind
And even if you decline
High fats and high stress
And you subscribe to exercise
GMOs could attest
To your eventual decline.
See, I know it's a mess
But before we hit rewind
Let's remind
ourselves
That behind our present
Wasn't any more pleasant
Due to the dollar signs

And anytime
This is the bottom line
of corporate slime,
With moral compasses misaligned
of execs in excess,
There's no success
That we can expect
Except
That it's we who possess
The ability
To feed the greed.
The responsibility lies
On both sides indeed.
And the seed of knowledge
While modified on one side
Au naturale on another
Must be mothered to rise
With a tempered hope
that it uncovers the lies
And the truth survives

In this piece Dwarka discusses processed foods, corporate control of food, advertising, genetically modified organisms, and health, among other points. Students should reference insights from the readings, as Dwarka does, and also be creative and demonstrate that they're thinking about cultural attitudes toward food and their own food choices.

Products and Assessment

The primary product for this lesson is the creative writing assignment and students' written response papers. These and students' participation in class discussion are the vehicles you can use to evaluate their achievement of lesson objectives.

I use the following grading criteria to evaluate student work for this lesson. I have, in different classes, formatted these criteria as a grading rubric or listed them in my course syllabus. How to format and apply the grading criteria will depend on individual context, school requirements, and teacher preference.

Grading Criteria

- Student demonstrates thoughtful reflection about personal food attitudes and practices.
- Student demonstrates knowledge of the features and issues associated with the modern food system, including issues such as industrial agriculture, food marketing, food and gender, globalization, genetically modified organisms, and food justice.
- Student demonstrates critical thinking about the role of culture in shaping attitudes toward food.
- Student demonstrates knowledge of the impact of the food system on other people, other beings, and the natural world.
- Student uses creative writing to reflect on personal experience and social phenomena.
- Student demonstrates critical analysis of the arguments made by authors, how they play out in their lived experiences, and their relevance to social and environmental justice.
- Work demonstrates thorough reading and comprehension of assigned course texts.
- Work demonstrates personal reflection, critical thought, and insight into course texts.
- Arguments are clear, well-developed, and documented with evidence from texts.
- Work demonstrates critical analysis of course topics, questions, and subject-matter.
- Style, usage, format, grammar, imagery, and presentation support meaning and are intentional, creative, and original.
- Work meets all requirements of the assignment and is utilized to facilitate development of personal understanding.

LESSON 4.2: FOOD JUSTICE

Access, Oppression, and Cruelty in Modern Food Systems

About this Lesson

In this lesson students learn about some injustices that are common in the U.S. within the modern food system. This includes the issue of food access and food deserts, and the ways that systemic lack of access to healthy food impacts certain populations. Students also learn about the experience of food workers, such as produce pickers and sprayers, who are often undocumented immigrants or other marginalized or oppressed groups. Finally, this lesson also addresses the cruel treatment that nonhuman animals experience in the modern food system, particularly within factory farms or confined animal feeding operations. Students consider the ethics of these practices and how conventional food consumption supports them.

Lesson Objectives

Students will:

- Gain knowledge of social and environmental justice issues associated with modern food production practices.
- Think critically about the ethical ramifications of modern food production and distribution.
- Gain knowledge of the impact of food production and consumption on marginalized human groups, other animals, and the land.
- Compose original analytical writing.
- Create original multi-media presentations.

Lesson Activities at a Glance

1. Students read about issues of justice relating to the modern food system.
2. Students write response papers outside of class.
3. Students discuss and reflect in class.
4. Students view documentary films outside of class and present reports on these films.

Key Content Area Skills: reading and analyzing nonfiction prose, writing analytical and reflective essays, analyzing documentary film, employing digital literacy to create multi-media presentations.

Texts Used in this Lesson

Philip Ackerman-Leist, *Rebuilding the Foodshed*, 135–158

Jeremy Rifkin, "Big Bad Beef," in *Reading the Environment*, 20–21

Barry Estabrook, *Tomatoland*, ix–34

Al Young, "Seeing Red," in *From Totems to Hip-Hop: A Multicultural Anthology of Poetry Across the Americas, 1900–2002*, 69

Kevin Bowen, "Gelatin Factory," in *Poetry Like Bread*, 68–69

Vandana Shiva, *Stolen Harvest*, 5–18

Peter Singer and Jim Mason, *The Way We Eat*, 3–55

Optional: Mark Winne, *Closing the Food Gap*, 21–34

Documentary Films

(Students sign up to view one outside of class)

> *The Garden*
>
> *Food, Inc.*
>
> *The World According to Monsanto*
>
> *Blue Gold*
>
> *King Corn*
>
> *The Future of Food*
>
> *Fresh*
>
> *Hungry for Change*
>
> *Forks over Knives*
>
> *Farmageddon*

Optional: *Earthlings* (be warned that this film is quite graphic)

Note: Publication information for these readings and films is listed at the end of the unit. Sample passages for readings are included in the full List of Readings at the end of the book.

My Procedure

To start, students read the assigned texts for this lesson and write response papers outside of class. These texts follow from those in Lesson 4.1; after students have been introduced to elements of the food system in that lesson, this lesson provides more detail about specific aspects of the food system that are unjust toward people, other animals, and the land.

The readings in this lesson often evoke anger and depression, or sometimes cynicism as students worry about the possibility of change. Students often become enraged at descriptions of the suffering endured to produce our food. For example, here's a student reflecting on the assigned selection from *Tomatoland* by Barry Estabrook – this reading illuminates negative aspects of the tomato-growing industry, including poor treatment of workers, heavy pesticide use, and growing and harvesting practices that are unhealthy for people and the land:

Student Writing: Response Papers

by Mallory Brooks

Excuse me, but is anyone else grossed out by the fact that they pick tomatoes when they are green and then keep them in a gas chamber to make them turn red? . . . The author made a point of the shape of the tomatoes produced on industrial farms, and it is true most tomatoes found in the supermarkets are completely round and taste like nothing. The ones out of my grandfather's garden are shaped like aliens and taste fabulous. . . . Whose bright idea was it to grow tomatoes in sand? Sometimes I think it's a great idea to light my hair on fire, and then I remember, it's actually a really dumb idea. Come on America, are we really that stupid? Don't answer that, I already know the answer. Clearly we are. The description of the way that tomatoes are grown in Florida was actually moronic to hear. Why in the world would we pump the ground with water and "chemotherapy" the ground? Seriously over 100 herbicides and pesticides. . . . Can we not do that? The Florida tomato production needs to shut it down. The attitudes of the companies selling these tomatoes need to also be shut down. "Taste does not enter the equation. 'No consumer tastes a tomato in the grocery store before buying it. I have not lost one sale due to taste,' one grower said. 'People just want something red to put on their salads." (Estabrook, 28). That is not cute. Shut it down you evil tomato man. To hear about the conditions that the workers are involved in even today is horrendous. "Workers were 'sold' to pay off bogus debts, beaten if they . . . were too sick or weak to work, held in chains, pistol whipped, locked at night into shacks in chain-link enclosures patrolled by guards. . . . " (Estabrook, XV). End tomato rant. (Although that poem was really sad, and now I want to free the tomatoes as badly as I want to free all the animals at the zoos. Lord help us all.)

Students are also struck by readings about the meat industry, including those by Rifkin, Bowen, and Singer and Mason, as they learn about the cruelty of factory farming facilities, the shear quantity of meat eaten in the U.S., and the environmental impact of raising large numbers of nonhuman animals to be killed for food. Here's an example of a student responding to Kevin Bowen's poem about the production of gelatin:

Student Writing: Response Papers

by Jennie Williams

[I]t was grotesque and shameful to read let alone understand the process of creating gelatin . . . It is so frightening to imagine piglets being raised only to be "boiled in acid, rendered a sticky mass rolled on screens and cooked in sheets to glass, smashed and ground to a fine power." All this terror is inflicted just for a simple sugary treat. . . . This poem opened my mind to really becoming more consciously aware of how and where processed items come from.

Students usually react strongly to this piece, often expressing significant empathy for the experience of the piglets who are killed to make gelatin and reflecting on how our society treats and thinks about these nonhuman animals. Here's another interesting comment on this piece:

Student Writing: Response Papers

by Chelsie Bateman

The production of gelatin is basically corpse desecration, unlike us burying our dead and our dead pets.

This very insightful comparison highlights the inconsistencies in our treatment of different species. After reading Bowen's poem, students often question whether they will ever eat gelatin again.

Note that this lesson isn't about advocating for or against eating certain foods, it's about thinking critically; I don't tell students to stop eating gelatin or meat or processed foods, though there are serious reasons to contemplate making such changes and many students do consider making at least some changes to their diet. But this lesson is about understanding the real story behind those foods – how they are made, where, with what ingredients, and who suffers along the way. Throughout this lesson, the key is for students to stare down the truth of what happens to other humans, other beings, and the land within our modern food system. Students should be thinking critically about why our food is grown in resource-intensive and pesticide-heavy monocultures. They should be thinking critically about why Americans eat as much meat as they do despite its huge environmental impact, when the same amount of grain that is used to feed animals raised for their flesh could feed many more people if fed directly to humans. They should ask why the animals raised for that meat are treated so horrifically, why so many Americans live in "food deserts" where they don't have access to healthy, fresh food, and why corporations are patenting seeds and gaining more and more control of the food supply.

Students often find the reality of these unhealthy and cruel practices confounding and simply absurd. Here's an example of a student pointing out this absurdity:

Student Writing: Response Papers

by Michelle Ott

There doesn't appear to be any logic behind what we are doing. We are wasting our lands, polluting our air, and further starving the poor to be able to provide unhealthy food to the rest of the population.

In addition to the readings in this lesson, I also have students watch documentary films about the food system outside of class and give presentations about these films. I have students sign up for one of the films, working in groups to present on this film in class. Here is the assignment:

Writing Assignment: Food Documentary Film Report

In class, you'll be given a list of films – sign up to view one of these films outside of class. Then, in small groups, prepare a report on the film to present to the class.
Your presentation should include the following components:

1. Describe the film. Identify the genre and year the film was made. Outline the plot or events and summarize key information communicated in the film – be specific, and consider using clips or images from the film to help your classmates get a sense of it.
2. What is the message of the film and what cultural discourses does it participate in or critique? Do the filmmakers have a specific opinion and point of view they are trying to convey? If so, what is it? How can you tell? Does the film have a 'moral'? How does the film approach issues we've discussed in class? Does it exemplify or challenge certain cultural attitudes surrounding food?
3. Your response to the film. Tell us what you thought of the film. Did you feel aligned with the perspective of the filmmaker? Did you identify with any of the characters? Why/why not? How did you feel after the film ended? Did it change your attitudes in any way?

Your presentation to the class should be 20 minutes in length. Submit your presentation with written notes and be sure to include full citation information at the end of the presentation for any sources cited in the presentation, including the film itself.

Students are generally very engaged by these films, and do a very effective job of sharing their analysis with the class. Including clips from the film and details about the events and messages portrayed in it helps give the rest of the class a good sense of some key ideas communicated in the film. Since these films apply to the ideas discussed in all three lessons in Unit 4, this assignment could be given to students in Lesson 4.1 or 4.3 as well. I typically give out the assignment early in the unit, give students time to view their films and prepare their presentations, and have them present later in the unit.

Products and Assessment

Participation in class discussion, written response papers, and in-class presentations on documentary films are the products you can use to evaluate students' achievement of the objectives for this lesson. Students should be thinking critically about the issues raised in this lesson, and while personal reflection is an important piece of that thinking, students must also demonstrate cultural critique that goes beyond their own habits or desires. (For example, defending "fast food" because they enjoy a particular fast food restaurant does not demonstrate a strong critique of how people and other animals are mistreated within this system.)

I use the following grading criteria to evaluate student work for this lesson. I have, in different classes, formatted these criteria as a grading rubric or listed them in my course syllabus. How to format and apply the grading criteria will depend on individual context, school requirements, and teacher preference.

Grading Criteria

- Student demonstrates knowledge of issues of justice associated with the modern food system, including access to healthy food, treatment of food workers, and treatment of nonhuman animals.
- Student demonstrates critical thinking about the impact of the food system on other people, other beings, and the natural world.
- Student demonstrates ethical reflection in considering the validity of modern food practices.
- Student demonstrates critical analysis of the arguments made by authors, how they play out in their lived experiences, and their relevance to social and environmental justice.
- Work demonstrates thorough reading and comprehension of assigned course texts.
- Work demonstrates personal reflection, critical thought, and insight into course texts.
- Arguments are clear, well-developed, and documented with evidence from texts.
- Work demonstrates critical analysis of course topics, questions, and subject-matter.
- Style, usage, format, grammar, imagery, and presentation support meaning and are intentional, creative, and original.
- Work meets all requirements of the assignment and is utilized to facilitate development of personal understanding.

LESSON 4.3: SUSTAINABLE FOOD MOVEMENTS

About this Lesson

In this lesson students explore proposals for sustainable alternatives to the modern food system, including calls for local and small farming, arguments for dietary changes, and discussions of permaculture, a method of growing food while supporting ecosystems and other beings. Students evaluate the potential of these alternative strategies for addressing the issues of justice and environmental degradation discussed in Lessons 4.1 and 4.2. They also locate and, if possible, visit a local community garden or farm to learn about the efforts of individuals in their community to participate in sustainable food practices.

Lesson Objectives

Students will:

- Gain knowledge of alternative modes of food production and consumption, including small-scale farming and permaculture.
- Think critically about the potential of alternative strategies to alleviate problems caused by modern industrial agriculture and the modern food system.
- Develop knowledge of local food growing efforts.
- Compose original analytical writing.

Lesson Activities at a Glance

1. Students read about strategies for food production that may be more environmentally sustainable and socially just than current practices.
2. Students write response papers outside of class.
3. Students discuss and reflect in class.
4. View film in class.
5. Students identify a local community garden or farm and visit this garden to interview a founder or worker (or interview by phone or email if a visit is not possible).

Key Content Area Skills: reading and analyzing nonfiction prose, writing analytical and reflective essays, conducting ethnographic interviews.

Texts Used in this Lesson

Wendell Berry, "The Pleasures of Eating," *Center for Ecoliteracy*

A. Breeze Harper, *Sistah Vegan*, 20–41, 80–86

Toby Hemenway, Gaia's Garden, 21–31, 120–123, 208–212

Rani Molla, "Can Organic Farming Counteract Carbon Emissions?" *Wall Street Journal*

UN Office of the High Commissioner for Human Rights, "Can Double Food Production in 10 Years, Says New UN Report," *United Nations Human Rights*
Optional: Barbara Kingsolver, *Animal, Vegetable, Miracle*

Film

Dirt! The Movie

Note: Publication information for these materials is listed at the end of the unit. Sample passages are included in the full List of Readings at the end of the book.

My Procedure

I start by having students read the assigned texts and write response papers outside of class. In this lesson students read recommendations for methods of food production and consumption that provide alternatives to modern industrial agriculture. These include calls to eat locally grown food, research into the environmental and economic value of small-scale organic farming, as well as permaculture, a method of "permanent agriculture" in which people design ecologically sound systems that grow food for humans while nourishing other animals and the land simultaneously. These strategies offer significant promise for providing greater food security and food access while supporting ecosystems.

It's extremely helpful to follow the bleak information in Lessons 4.1 and 4.2 with the encouraging possibilities detailed in this lesson. Students find inspiration in these ideas, and connect this inspiration to their own growing desires to eat in healthier and more sustainable ways. One student says:

Student Writing: Response Papers

by Mallory Brooks

Have you ever had a fresh tomato, that someone locally grew, perhaps even out of a community garden? Its the best daggone thing you'll ever taste in terms of tomatoes. I think it's so important to support the little farms because those are they [*sic*] people like you and me, who will work from sun up to sun down to get their crop on to someone's table. The big farms don't care about quality, it's all about quantity. They will paint their fields with poison in fear of losing a dollar before they would ensure that what they are producing actually tastes good.

When students come to class after reading these materials, I start by having them talk about the readings in small groups. I ask them to discuss which of the strategies they've read about they feel has the most promise for changing the problems they've learned about in Lessons 4.1 and 4.2. Each member of the group should give their opinion as to which strategies they find most promising, and support this opinion with reasoning as to why they think that strategy has the best chance of making positive change. When they've finished discussing in small groups, I ask each group to recap their discussion for the class, and we discuss students' opinions as a whole group.

In class we also watch the film *Dirt!*, which offers a great comprehensive exploration of what modern practices have done to the soil and the importance of regenerating soil nutrients and micro-organisms. Students respond very positively to this film, generally loving the insights and examples it provides.

Next, to connect these ideas to students' own local communities, I have students locate a small farm or community garden in their area and learn about its practices. In my own classes, this means learning about small urban farms and community gardens in and around Baltimore City. I have students work in groups and then have each student add a section of writing to a shared Wiki within our online course site. However, this assignment could be structured with individual written reports, presentations, or other methods of communicating what students have learned. Here is my assignment:

Assignment: Local Farm or Community Garden Visit

Working in groups, arrange a visit to a local area farm or community garden. Contact the proprietors in advance to set up the visit, get a tour, and talk to a representative of the farm about their endeavor. Ask them about their farm (How long has it been in operation? What are their goals? How many people work or volunteer on the farm? What items does the farm produce? What methods of growing do they use and how do they seek to support the environment and their community?).

Then, individually, you will contribute your observations and thoughts about this visit via the Wiki page in our online course site. Each person will add at least 500 words of their own text to a common page for each farm/garden.

Here's an excerpt from one student's write-up of his farm visit:

Student Writing: Farm Visit

by Danny Clemens

What struck me most about [the] farm was how deeply entrenched the farm was within its community - on the Sunday evening in October that we visited, the farm was bustling with families visiting the pumpkin patch . . . and buying fresh produce. Where I live . . . there are few (if any) farms that continue to operate with such a strong community presence. Although there are local farms that are popular for hay rides and pumpkin picking around Halloween, [this farm] is an institution all season long and attracts a great deal of business.

Sadly, the greenhouse was closed to the public so we were not able to tour it but I was able to see a variety of flowers, bushes and small trees growing inside. I think this speaks to [the farm's] commitment to sustainable agriculture: they use their land to cultivate all different types of flora and fauna. Their diverse harvest is not only fun to tour but also beneficial for the soil and the ecosystems that live in the farm environment.

Students find great value in the ideas they read about in this lesson, and in their visits to local farms and community gardens. These experiences allow them to recognize positive alternatives to the modern industrial food system that they've learned is so problematic. One student writes:

Student Writing: Response Papers

by Yiannis Balanos

It is very easy to lose touch from the source of food, but on the most fundamental level, we all originate and exist directly from the sun and soil. In order to prevent this kind of disconnect, I think it is important to teach children about food and its sources. At the same time, it is easy to fall from this connection growing older. Thus, it would greatly benefit society to fund field trips to farms for middle and high school students, promote conscious eating, and having locally grown food as available as possible.

Products and Assessment

Participation in class discussion, written response papers, and students' write-ups of their farm visits are the products you can use to evaluate students' achievement of the objectives for this lesson. Students should be thinking critically about the possible benefits and implications of the alternative methods of food production they've read about.

I use the following grading criteria to evaluate student work for this lesson. I have, in different classes, formatted these criteria as a grading rubric or listed them in my course syllabus. How to format and apply the grading criteria will depend on individual context, school requirements, and teacher preference.

Grading Criteria

- Student demonstrates knowledge of proposed alternative methods of food production.
- Student demonstrates thoughtful reflection about the benefits and possible outcomes of alternative methods of food production.
- Student demonstrates critical thinking about how alternative methods of food production and consumption could address issues with modern industrial agriculture, such as food security, food access, cruelty, and environmental degradation.
- Student demonstrates increased knowledge of local food production operations and critical thinking about the benefit of these operations.
- Student demonstrates critical analysis of the arguments made by authors, how they play out in their lived experiences, and their relevance to social and environmental justice.

- Work demonstrates thorough reading and comprehension of assigned course texts.
- Work demonstrates personal reflection, critical thought, and insight into course texts.
- Arguments are clear, well-developed, and documented with evidence from texts.
- Work demonstrates critical analysis of course topics, questions, and subject-matter.
- Style, usage, format, grammar, imagery, and presentation support meaning and are intentional, creative, and original.
- Work meets all requirements of the assignment and is utilized to facilitate development of personal understanding.

EFFECT OF THE LESSONS: PERSONAL AND SYSTEMIC IMPACTS, INTERDEPENDENCE, AND GLOBALIZATION

The study of food connects our most personal choices to the most global and far-reaching systems and effects. The consequences of what we each consume extend far beyond our own bodies, affecting not only our local communities and ecosystems but also people, nonhuman animals, and ecosystems we have never seen directly in distant parts of the world. By learning about the modern food system, students can vastly increase their understanding of interconnected social justice and environmental issues. They can also gain insight into global economic and political dynamics, and see the connections between these abstract systems and their own practices, including connections to what food is available to them, what growers and producers they support through their eating choices, and how their health and the health of others is affected. Studying food systems is also an essential time to look closely at the rights of other living beings, and consider our treatment of and attitudes toward these beings. What does it mean for plants, animals, soils, and ecosystems if we choose to grow large monoculture crops that require heavy pesticide use and land clearing? What does it mean if we genetically modify and claim corporate ownership of seeds? If we subject animals to stunningly inhumane treatment in order to satisfy a culturally fueled demand for more and more meat? If we clear rainforest to produce grain that will be used, not to feed people, but to raise animals who will then be killed to feed far fewer people?

One key element that these lessons should provide for students is a more nuanced understanding of these interlocking global issues and forces. For example, students should start to comment on ramifications of government policies like subsidies, as one of my students does here relating to beef consumption:

Student Writing: Response Papers

by Chelsie Bateman

The government is supporting our huge consumption by backing ranchers with almost free cattle to raise. So, we need the government to view our predicament more like a chain reaction because that is what it essentially is; a chain reaction of events that eventually comes back to affect us.

Here's another example in which a student explores complex causes and effects in relation to problems with the food system:

Student Writing: Response Papers

by Yiannis Balanos

In *Rebuilding the Foodshed*, the reading expressed some attitudes about the food system that I had not heard before. It states that revitalizing local food economies and traditions can provide opportunities against disenfranchisement and a strong voice. This [*sic*] definitely true as many indigenous people, among other minorities, have lost rights and are forced to grow produce that will eventually be sold back to them. Such injustices that can be solved with a bottom-top approach have been ignored ultimately because of the control that large industries possess. If people are given the power to support themselves and manage farms individually, many injustices that are brought by large business can be eliminated. I believe that ultimately, since agriculture is such a vital part of human survival, farms in the hands of individuals will be at least as, if not more, successful than one giant company. . . . Furthermore, since farmers take pride in their work, land, and crop, giving them control of farms will make them happier, eliminating work maltreatment. We have been in a monoculture and large business mindset for too long.

As students think about these complex systems, they should also reflect closely on their own practices and attitudes, and on the very personal connections that food can help us forge with other beings. Food is a powerful way to remember our interdependence, as we cannot survive without soil, plants, and the myriad organisms who work within and around the soil and plants to help our food grow. Students should come to recognize the importance of respecting these connections, being aware of how our food is produced, and eating in ways that are ethical and support justice:

Student Writing: Response Papers

by Yiannis Balanos

Growing food, something so essential to life and having such a deep connection with the earth, should be ethical. Otherwise, the way we treat the earth reflects on us and exposes us for who we really are as a race.

During this unit many students point out that our disconnection from other people and other beings allows us to ignore their suffering. Here's an example of a great insight from an anonymous student:

Student Writing: Response Papers

I feel that the disconnect between the animal products that we use and the animals themselves contributes to this problem. If we knew where our food came from and knew the animals that were slaughtered for our food, I believe that we would be much more likely to appreciate and respect non-human life.

Finally, the lessons in this unit should also encourage students to think about different possibilities for producing our food – ways that do not support the model of industrial agriculture and instead build ecological and social health and community. One student comments:

Student Writing: Response Papers

by Dwarka Pazavelil

I feel more connected to my community in particular and this sense of partnership makes it easier to take the next steps to eat for my own health as well as the health of the environment.

UNIT 4 READINGS AND REFERENCES

Ackerman-Leist, Philip. 2013. *Rebuilding the Foodshed: How to Create Local, Sustainable, and Secure Food Systems.* White River Junction, VT: Chelsea Green Publishing.

Benenson, Bill, Gene Rosow, and Eleonore Dailly. 2009. *Dirt! The Movie.* Documentary. Common Ground Media.

Berry, Wendell. 2013. "The Pleasures of Eating." *Center for Ecoliteracy.* Accessed August 29. www.ecoliteracy.org/essays/pleasures-eating.

Bowen, Kevin. 2001. "Gelatin Factory." In *Poetry Like Bread*, edited by Martín Espada, 68–69. Willimantic, CT: Curbstone Press.

Bozzo, Sam. 2010. *Blue Gold: World Water Wars.* Documentary. Purple Turtle Films.

Canty, Kristin. 2011. *Farmageddon.* Documentary. Kristin Canty Productions.

Colquhoun, James, Laurentine Ten Bosch, and Carlo Ledesma. 2012. *Hungry for Change.* Documentary.

Estabrook, Barry. 2012. *Tomatoland: How Modern Industrial Agriculture Destroyed Our Most Alluring Fruit.* Reprint ed. Andrews McMeel Publishing, LLC.

Fulkerson, Lee. 2013. *Forks Over Knives.* Documentary. Monica Beach Media.

Harper, A. Breeze. 2010. *Sistah Vegan: Food, Identity, Health, and Society: Black Female Vegans Speak.* New York: Lantern Books.

Hauter, Wenonah. 2012. *Foodopoly: The Battle Over the Future of Food and Farming in America.* Reprint ed. New York: New Press.

Hemenway, Toby. 2009. *Gaia's Garden: A Guide to Home-Scale Permaculture.* 2nd ed. White River Junction, VT: Chelsea Green Publishing.

Higman, B. W. 2011. *How Food Made History.* 1st ed. Chichester, West Sussex, UK and Malden, MA: Wiley-Blackwell.

Joanes, Ana Sofia. 2009. *Fresh*. Documentary. Ripple Effect Films.

Kennedy, Scott Hamilton. 2014. *The Garden*. Documentary. Black Valley Films.

Kenner, Robert. 2008. *Food, Inc.* Documentary. Magnolia Pictures.

Kingsolver, Barbara. 2007. *Animal, Vegetable, Miracle: A Year of Food Life*. New York: HarperCollins.

Koons, Deborah. 2004. *The Future of Food*. Documentary. Lily Films.

Lappé, Anna, and Bryant Terry. 2006. *Grub: Ideas for an Urban Organic Kitchen*. New York: Tarcher.

Levenstein, Harvey A. 1993. *Paradox of Plenty: A Social History of Eating in Modern America*. New York: Oxford University Press.

Molla, Rani. 2014. "Can Organic Farming Counteract Carbon Emissions?" *Wall Street Journal: The Numbers*. May 22. http://blogs.wsj.com/numbers/can-organic-farming-counteract-carbon-emissions-1373/.

Monson, Shaun. 2005. *Earthlings*. Documentary. Nation Earth.

Parasecoli, Fabio. 2008. *Bite Me: Food in Popular Culture*. New York: Berg.

Parkin, Katherine. 2000. "Campbell's Soup and the Long Shelf Life of Traditional Gender Roles." In *Kitchen Culture in America: Popular Representations of Food, Gender, and Race*, edited by Sherrie A. Inness, 51–64. Philadelphia: University of Pennsylvania Press.

Patel, Raj. 2008. *Stuffed and Starved: The Hidden Battle for the World Food System*. Brooklyn, NY: Melville House.

Pollan, Michael. 2007. *The Omnivore's Dilemma: A Natural History of Four Meals*. New York: Penguin.

Rifkin, Jeremy. 1994. "Big, Bad Beef." In *Reading the Environment*, edited by Melissa Walker, 1st ed., 20–21. New York: W. W. Norton.

Robin, Marie-Monique. 2008. *The World According to Monsanto*. Documentary. Image et Compagnie.

Root, Waverly, and Richard De Rochemont. 1981. *Eating In America*. Hopewell, NJ: Ecco.

Ryan, John C., and Alan Thein Durning. 1997. *Stuff: The Secret Lives of Everyday Things*. Seattle, WA: Northwest Environment Watch.

Shiva, Vandana. 2000. *Stolen Harvest: The Hijacking of the Global Food Supply*. Cambridge, MA: South End Press.

Singer, Peter, and Jim Mason. 2006. *The Way We Eat: Why Our Food Choices Matter*. Emmaus, PA: Rodale.

United Nations Office of the High Commissioner for Human Rights. 2011. "Can Double Food Production in 10 Years, Says New UN Report." *United Nations Human Rights*. March 8. www.ohchr.org/EN/NewsEvents/Pages/DisplayNews.aspx?NewsID=10819&LangID=E.

Winne, Mark. 2009. *Closing the Food Gap: Resetting the Table in the Land of Plenty*. Reprint ed. Boston: Beacon Press.

Woolf, Aaron. 2007. *King Corn*. Documentary. ITVS.

Young, Al. 2003. "Seeing Red." In *From Totems to Hip-Hop: A Multicultural Anthology of Poetry Across the Americas, 1900–2002*, edited by Ishmael Reed, 69. New York: Thunder's Mouth Press.

Unit 5

STUFF: PRODUCTION, CONSUMPTION, AND WASTE

Imagine what you would come upon today at a typical landfill: old furniture, upholstery, carpets, televisions, clothing, shoes, telephones, computers, complex products, and plastic packaging, as well as organic materials like diapers, paper, wood, and food wastes. Most of these products were made from valuable materials that required effort and expense to extract and make, billions of dollars' worth of material assets. The biodegradable materials such as food matter and paper actually have value too – they could decompose and return biological nutrients to the soil. Unfortunately, all of these things are heaped in a landfill, where their value is wasted. They are the ultimate products of an industrial system that is designed on a linear, one-way cradle-to-grave model. Resources are extracted, shaped into products, sold, and eventually disposed of in a "grave" of some kind, usually a landfill or incinerator. You are probably familiar with the end of this process because you, the customer, are responsible for dealing with its detritus. Think about it: you may be referred to as a consumer, but there is very little that you actually consume – some food, some liquids. Everything else is designed for you to throw away when you are finished with it. But where is "away"? Of course, "away" does not really exist.

—*William McDonough and Michael Braungart (27)*
Lesson 5.1

The lessons in this unit focus on how consumer goods are made, how and why we consume these goods as we do, and what happens to them once we dispose of them. Students will learn about the many resources that go into making our "stuff" and the environmental and social impacts of this production and consumption.

They will also be asked to consider the cultural factors that influence consumption and reflect thoughtfully on what they themselves buy and why they buy it.

Some key questions that the lessons in this unit should raise for students:

- What consumer goods do you buy on a regular basis? What consumer goods do you need? What goods do you want? Why do you want these?
- What messages does our culture send us about buying consumer products?
- What processes and materials are involved in making our consumer goods? How does the making of these products affect people, other animals, and the land?
- How can we buy, make, and dispose of goods differently to achieve more sustainable conditions?

LESSON 5.1: ABOUT OUR STUFF

How Things Are Made and Consumed in the U.S.

About this Lesson

In this lesson students explore the processes and resources involved in making consumer goods, how those goods are distributed and consumed, and the social and environmental impact of this production and consumption.

Lesson Objectives

Students will:

- Gain knowledge of how consumer goods are produced, consumed, and disposed of.
- Gain knowledge of the social and environmental impact of modern production, consumption, and waste.
- Reflect on their own consumption of consumer goods and the hidden impact of this consumption.
- Compose original analytical writing.

Lesson Activities at a Glance:

1. Students read assigned texts about methods of production and the scope of modern consumption and waste.
2. Students write response papers outside of class.
3. Students watch video and view photography in class.
4. In-class discussion.

Key Content Area Skills: reading and analyzing nonfiction prose, writing analytical and reflective essays.

Texts Used in this Lesson

John Ryan and Alan Thein Durning, *Stuff: The Secret Lives of Everyday Things,* 7–25, 62–66

Mathis Wackernagel and Williams Rees, *Our Ecological Footprint: Reducing Human Impact on the Earth,* 7–30

Ronald Wright, *A Short History of Progress,* 29–65

William McDonough and Michael Braungart, *Cradle to Cradle: Remaking the Way We Make Things,* 17–37, 92–102

Rita Turner, "Discourses of Consumption in US-American Culture," *Sustainability*

David Orr, "The Carbon Connection," *Center for Ecoliteracy*

Jim Merkel, *Radical Simplicity*, 2–16

Andrew Marshall, "No Conspiracy Theory – A Small Group of Companies Have Enormous Power over the World," *AlterNet*

Alan Thein Durning, "The Dubious Rewards of Consumption," in *The Earthscan Reader in Sustainable Consumption*, 129–134

Optional: Thomas Princen, "Distancing: Consumption and the Severing of Feedback," in *Confronting Consumption*, 103–131

Optional: Jack Manno, "Commoditization: Consumption, Efficiency and an Economy of Care and Connection," in *Confronting Consumption*, 67–99

Optional: Bill McKibben, *Deep Economy*, 5–45

Video

The Story of Stuff, produced by the Story of Stuff Project with Annie Leonard

Trailer for the film *Midway* by Midway Film and Chris Jordan

Photography

Works of Chris Jordan: "Intolerable Beauty: Portraits of American Mass Consumption" and "Midway: Message from the Gyre"

Note: Publication information for these materials is listed at the end of the unit. Sample passages are included in the full List of Readings at the end of the book.

My Procedure

Students read the assigned texts for this lesson outside of class, and write response papers that they post online. The texts in this lesson explore the details and effects of modern production processes (in pieces like Ryan and Durning, Rees et al., and McDonough and Braungart), as well as critiques of consumer culture (in pieces like Merkel, Turner, and Durning). This is a lot for students to take in, but typically evokes passionate reactions.

One of the pieces students respond most strongly to is *Stuff* by Ryan and Durning. This text traces the details of how everyday products, like coffee and T-shirts, are produced. Students are usually shocked to learn of the impacts of products they regularly use, and quickly begin to reflect on their own consumption. Here's an example of a student discussing these new realizations:

Student Writing: Response Papers

by Jennie Williams

When reading the section on coffee, I was forced to reflect on my own consumption of the brown colored, breath tainting liquid that is just so good! It is amazing to think that because of my two cups a day, I have 12 coffee trees growing per year all for me. Which means my mom has 24 coffee trees devoted to her. It is so strange to put it into perspective because the common thought is how could any one individual make such an impact?

As students are exposed to impacts of modern consumer goods, they also explore critiques of our methods of production and consumption. Here an anonymous student discusses McDonough and Braungart's text *Cradle to Cradle*:

Student Writing: Response Papers

According to Cradle to Cradle, life functions in a continuous cycle where energy flows from one source to another. Unfortunately, people have been consuming and throwing away "monstrous hybrids" of biological and technical mass that disrupt the equilibrium of materials (98). These hybrids are neither salvageable nor recyclable leaving them as wasted lost items that build mountains of trash. The authors purpose [*sic*] a hopeful message to the audience that society can change if we adapt to a mentality of living "cradle to cradle" verses "cradle to grave" (27).

When students come to class, we begin by discussing their reactions to these texts. This includes talking through the feelings of surprise and anger that students often experience when learning about the harmful effects of consumer products. To enhance students' understanding of these issues, in class I show the video *The Story of Stuff*. This excellent 20-minute video provides a clear and engaging outline of many of the topics covered in the readings for this lesson, and helps students connect the ideas together.

In class we also view and discuss the photography of Chris Jordan. Along with the assigned readings, I assign students to view some of Jordan's photo series outside of class, as well as a video about Jordan's work titled *Midway*. In class, we pull up some of Jordan's photographs and discuss students' reactions. Jordan's work includes photographs of the sheer mass of products disposed of in landfills, as well as a series in which he captures images of birds who died from eating plastic (which they see floating in oceans and mistake for food). Here's an example of a student commenting on Jordan:

Student Writing: Response Papers

by Mallory Brooks

After looking at the "readings" (and video), I have to admit, I'm feeling a bit emotional. I have recently been trying to be more eco friendly . . . it isn't easy, but the effort feels worth it. I have to admit, I still do lean on grabbing bottled water if I'm in a hurry and my water canister is upstairs, but after seeing the trailer for Midway, I will be avoiding these moments of laziness and dashing for the stairs. Now, most of you know, I talk . . . a lot. I share how I feel, and at this point I'm at a loss for words after seeing what these birds have been dying from. I cried. I am an animal lover. In fact, I have caught a wild duck that had hurt his foot, drove him to an exotic vet and had him checked out. That is a story for another time, but I know if I went to this island, I would be at a loss. What can be done for the hundreds, maybe thousands of birds who are suffering at the cost of our plastic bottles, and other non-degradable items that have been carelessly tossed on the ground or in the ocean? The answer seems to be nothing . . . But there has to be something. I'm feeling a bit down and have little to say about what I have read, and what I have seen. I feel almost ashamed to be part of humanity that has slowly, but surely worked so hard for our convenience at the cost of our world. This is not going to be an easy class to sit through, I feel like its going to be really challenging to try to understand why our habits are so destructive. It seems like . . . so much for one person to try to change, and it is, but there has to be others who are feeling in a similar way, right?

Like some other lessons in this book, in-class discussion of the readings for this lesson should involve helping students work through their feelings of anger and grief over the impacts of our everyday actions. This is an opportunity to critique social pressures to consume. Some authors assigned for this lesson propose significantly reducing our individual consumption, and this is an important possibility for students to consider, but students should also consider changing *how* we make and "dispose of" the goods we use.

Products and Assessment

Participation in class discussion and written response papers are the products you can use to evaluate students' achievement of the objectives for this lesson. Students should be demonstrating new understanding of the ramifications of our modern processes of production, consumption, marketing, distribution, and waste of consumer goods, and linking these understandings to thoughts about their own behavior.

I use the following grading criteria to evaluate student work for this lesson. I have, in different classes, formatted these criteria as a grading rubric or listed them in my course syllabus. How to format and apply the grading criteria will depend on individual context, school requirements, and teacher preference.

Grading Criteria

- Student demonstrates knowledge of the scope and methods of modern production, consumption, and waste.
- Student demonstrates thoughtful reflection on the social dimensions and motivations behind modern consumption.
- Student demonstrates knowledge of the social and environmental impact of production, consumption, and waste.
- Student demonstrates personal reflection on her/his own consumption and its social and environmental impact.
- Student demonstrates critical analysis of the arguments made by authors, how they play out in their lived experiences, and their relevance to social and environmental justice.
- Work demonstrates thorough reading and comprehension of assigned course texts.
- Work demonstrates personal reflection, critical thought, and insight into course texts.
- Arguments are clear, well-developed, and documented with evidence from texts.
- Work demonstrates critical analysis of course topics, questions, and subject-matter.
- Style, usage, format, grammar, imagery, and presentation support meaning and are intentional, creative, and original.
- Work meets all requirements of the assignment and is utilized to facilitate development of personal understanding.

LESSON 5.2: FOLLOW THAT STUFF

Profile of a Product

About this Lesson

In this lesson students select a common consumer product and research how and where it is produced, what materials are used to make it, and the environmental impact of its production, shipping, and disposal.

Lesson Objectives

Students will:

- Gain knowledge of how consumer goods are produced, consumed, and disposed of.
- Gain knowledge of the social and environmental impact of modern production, consumption, and waste.
- Reflect on their own consumption of consumer goods and the hidden impact of this consumption.
- Compose original analytical writing.

Lesson Activities at a Glance:

1. Students work in groups to select consumer products and research their production.
2. Groups share their research with the class.

Key Content Area Skills: conducting original research.

Texts Used in this Lesson

There are no new readings for this lesson. Refer back to Ryan and Durning's *Stuff: The Secret Lives of Everyday Things* from Lesson 5.1.

My Procedure

This lesson refers to the reading by Ryan and Durning from Lesson 5.1, *Stuff: The Secret Lives of Everyday Things* – review that text with students if necessary. To start this lesson, I put students in groups and ask them each to select a common consumer object – in the past I've had groups select toothpaste, mattresses, pencils, diamond rings, and a popular brand of scented body spray for men. Once students have selected their object, I charge them with researching how and where that object is produced. I ask them to model this research on *Stuff:*

The Secret Lives of Everyday Things; while I don't expect them to find information that is as thorough or in-depth as the research in that book, I want them to aim to find out the same sorts of things that are presented about each object in that text. This should include, as much as they can find: what materials are used in making the product, where those materials are acquired/mined/harvested, where the materials are shipped to be processed, the steps involved in making the product itself, any byproducts of the manufacturing process, how far the product would be shipped to be sold in the students' area, what is done with the product after it is used (including the packaging), and if the product is disposed, the environmental impacts of its disposal.

In most of my classes I have students do this research during the class period, making sure that at least one member of each group has brought a laptop or tablet that they can use to do internet research and moving between the groups myself to support and guide their searches. I have them research together throughout the class period, and at the beginning of the next class they share all of their findings with the rest of the class, and submit to me a list of their references (I have students compile this list of references as they're working and email it to me when they've finished their research).

However, I have also taught classes where I've made this a more extended activity and had students research their product outside of class for part of the semester, putting together a formal presentation to share their findings. For this version I provide a list of objects for students to pick from, although I allow groups to propose alternatives. The list includes plastic water bottles, diamond rings, athletic shoes, denim jeans, Barbie dolls, high-heeled shoes, television sets, lipstick, and disposable razors. For this longer assignment I also ask students to look at the cultural background of the object and how it is marketed, in addition to researching how it is made, distributed, and disposed of. The following box contains the longer version of the assignment.

Assignment: Object Case Study

In groups, you will select an object (a list of options will be provided in class) and prepare a presentation outlining this object's history and production and analyzing its cultural significance. After your group conducts its in-depth research on this object, you will lead a class session in which you present information and facilitate class discussion about the object.

Your case study presentation will consist of several components:

1. Research the history of the object, the role it has played in society, and how it has been advertised over time (as part of your presentation on the object's history, you may wish to share one historical advertisement and one modern advertisement).

(continued)

(continued)

2. Research the modern production chain of this object. Find out what materials this object is made from today, where those materials are sourced/grown/mined, where the materials are shipped to be used in production, what processes are used during production, the byproducts of production, how far the object is commonly shipped for distribution, and what is typically done with the object once it is no longer in use. Consider the social and environmental impacts of each step in this process.
3. Share your own analysis of the object, thinking about how it is used by people, what its social significance is, and what cultural groups are most likely to use this object and why. Then lead the class in discussion about this object. Prepare at least ten discussion questions to use as you facilitate this conversation.

For your presentation, you must submit both a presentation file (such as PowerPoint) and a narrative of what you will say in class, including notes outlining your research, analysis, and full citations for all sources.

Since students give these presentations live in class, I won't be quoting from examples of their work here. Students almost always turn up very interesting and disturbing information during their research, finding out things they didn't know, such as unexpected chemicals in personal toiletry products, toxic pollution caused during production, poor labor conditions for workers in factories, animal testing conducted on products, and more. Sharing the results of this research is generally very engaging for students, and gives them far greater insight into the typical processes behind products they use every day.

Products and Assessment

The primary product that comes from this lesson is students' reports on their research findings and their lists of references. For the shorter version of this assignment I don't usually require a written document to accompany students' in-class reports (other than their references); however, when I make this a longer assignment with a more formal presentation I do require a written document in addition to a PowerPoint or other presentation file.

I use the following grading criteria to evaluate student work for this lesson. I have, in different classes, formatted these criteria as a grading rubric or listed them in my course syllabus. How to format and apply the grading criteria will depend on individual context, school requirements, and teacher preference.

Grading Criteria

*   Student conducts original research using reliable and documented sources.
*   Student demonstrates knowledge of the scope and methods of production, consumption, and waste for a particular consumer product.

- Student demonstrates thoughtful reflection on the social dimensions and motivations behind modern consumption.
- Student demonstrates knowledge of the social and environmental impact of production, consumption, and waste for a particular consumer product.
- Student demonstrates personal reflection on her/his own consumption and its social and environmental impact.
- Work demonstrates thorough reading and comprehension of assigned course texts.
- Work demonstrates personal reflection, critical thought, and insight into course texts.
- Arguments are clear, well-developed, and documented with evidence from texts.
- Work demonstrates critical analysis of course topics, questions, and subject-matter.
- Style, usage, format, grammar, imagery, and presentation support meaning and are intentional, creative, and original.
- Work meets all requirements of the assignment and is utilized to facilitate development of personal understanding.

LESSON 5.3: HOW OUR CONSUMPTION AFFECTS OTHERS

About this Lesson

In this lesson students reflect on what they have learned in Lessons 5.1 and 5.2 about modern production and consumption of goods by writing poetry and short stories exploring the experiences of nonhuman beings who are affected by the production of a particular consumer good.

Lesson Objectives

Students will:

- Reflect on the impact of producing consumer goods.
- Use creative writing to imagine the experiences of others.
- Reflect on their own consumption of consumer goods and the hidden impact of this consumption.
- Compose original poetry or creative prose.

Lesson Activities at a Glance

1. In-class discussion.
2. Students complete a creative writing assignment reflecting on other beings affected by the production of a particular product.
3. Students share creative writing in class.

Key Content Area Skills: writing poetry and creative prose.

Texts Used in this Lesson

This lesson has no new assigned texts, but builds on those discussed in Lesson 5.1.

My Procedure

I start this lesson with class discussion, reviewing with students some of the impacts of consumer goods they have learned about from Lessons 5.1 and 5.2. I ask them to share some of the biggest surprises from their reading and research in those lessons, and to list some of the many types of environmental impacts caused by consumer products.

Next I have students brainstorm a list of nonhuman beings who are likely affected by these many types of environmental impacts. If producing a product causes pollution runoff in a local stream, what creatures are affected? If mining materials used to make a product strips a local ecosystem, what creatures are affected? If a product is tested on nonhuman animals, who are those animals who are being tested on?

This list is intended to encourage students to think about the lives of other beings touched by the harmful processes of production discussed in this unit.

I next introduce the following creative writing assignment:

Writing Assignment

Write a poem or short story from the perspective of a nonhuman being who is connected to the process of producing a product or material you use in your daily life. This may be a tree cut down to make paper you write on, a mountain from which coal is extracted that powers your computer, a fish in a river that contains runoff from the production of a product you use, a plant or nonhuman animal that is raised for food which you then eat, an insect living on a plant which is being grown for food, etc. Think about the details of how this creature is affected by the production and consumption of the specific product or material, and think about how this creature experiences these effects.

I do allow students to consider beings affected by the production of food for this assignment, in addition to considering other consumer goods. The line between edible consumer products and other types of products is rather blurred in our contemporary mass-marketed consumer culture, and students have read about consumables like coffee and soda in the reading by Ryan and Durning for Lesson 5.1. Modern food products are often produced, marketed, and shipped long distances very much like other consumer goods, and are often sold with as much packaging (using materials like plastic and aluminum). There are many beings affected by these production methods along the way, so the assignment can work well when applied to food products.

Below is an example of a very thoughtful poem written for this assignment.

Student Writing: Creative Writing Assignment

by Heather Harshbarger

"Milking"
Eventually, we all fall prey
to the ultimate predator machine.
Cut me into a succulent filet,
take away all life's gleam,

(continued)

(continued)

> stuff me with nutritional landfill.
> I'll chew on yesterday's meal
> and stay in my penned mill.
> No, I don't think you know I feel
> the stab of the blade, the brand;
> this isn't home, I'm in demand
> for tomorrow's best-cut steak,
> or I'd be dead by corn long before
> it really should take
> one cow to hit the floor.
> Tainted with chemicals, poisons, pesticide,
> my fat transfers to yours, glowing,
> golden, modified; fine-cut joke, inside
> me is a wasteland that's growing
> as large and as deadening, as cunning,
> as the lives this place has been numbing.

When students come to class after completing this assignment, I ask if any are comfortable volunteering to share what they wrote. I also ask them what creatures they chose to write about and what insights they gained from doing this assignment.

What I look for in this assignment is evidence that students are applying some of the information they have learned about production, consumption, and disposal of consumer merchandise, and that they are demonstrating genuine empathy as they think about the impact of what we buy, make, and dispose of on other beings. I want to see specific details that illustrate the effects of production, shipping, or disposal on the being they are considering. Many students do choose to write about animals affected by food production, specifically those in factory farms. The sort of cruelty inflicted in factory farms often weighs heavily on students' minds, as it should, and this assignment is an opportunity for students to express that.

Products and Assessment

The product of this lesson is students' creative writing. This can be used to evaluate their achievement of the lesson objectives.

I use the following grading criteria to evaluate student work for this lesson. I have, in different classes, formatted these criteria as a grading rubric or listed them in my course syllabus. How to format and apply the grading criteria will depend on individual context, school requirements, and teacher preference.

Grading Criteria

- Student demonstrates knowledge of the scope and methods of modern production, consumption, and waste.

- Student demonstrates thoughtful reflection on the impact that production, consumption, and waste has on other beings.
- Student thoughtfully imagines the lives of nonhuman creatures.
- Student demonstrates empathy in imagining the experiences, feelings, and desires of others.
- Work demonstrates thorough reading and comprehension of assigned course texts.
- Work demonstrates personal reflection, critical thought, and insight into course texts.
- Arguments are clear, well-developed, and documented with evidence from texts.
- Work demonstrates critical analysis of course topics, questions, and subject-matter.
- Style, usage, format, grammar, imagery, and presentation support meaning and are intentional, creative, and original.
- Work meets all requirements of the assignment and is utilized to facilitate development of personal understanding.

EFFECT OF THE LESSONS: CONSUMER CULTURE AND IMPACTS ON ECOLOGICAL SYSTEMS

An important outcome of the lessons in Unit 5 is to get students thinking about modern consumer culture. Mainstream culture sends out near-constant messages encouraging us to buy things. It's important for students to think about why this is and what effect it has, on us, on our communities, and on the natural world. These things we buy don't exist in a vacuum – they're made of materials that are carved from the earth or taken from other beings, put through manufacturing processes that may cause enormous side effects, and shipped long distances to reach the hands of the person we've deemed the "consumer." Such practices shouldn't be taken lightly, and buying consumer goods should be done with full awareness of those affected.

Through engaging with the lessons in this unit, students should start to realize this fact, that all of our "stuff" is made from precious materials and lives – that we mine, cut, harvest, and kill, as well as pollute and exploit workers, in order to get these objects. It's all too easy for those consuming not to think about the lives their products are built on and from – here's an example of an anonymous student pointing this out:

Student Writing: Response Papers

In our society today we often forget that everything we buy came from somewhere and had to be assembled, grown, or built. This is a move for the worse. If we truly want to live in harmony with the environment, we should understand these connections and the impacts of the goods we buy. The less that we know about where our goods come from, the greater that our environmental impact will be, resulting in negative changes in the world around us.

As students think about the origins of their products, they should also start to think about *why* we buy so many material goods in the U.S. Many students come to realize that the reason we feel we have to buy so much is largely cultural. Dominant society has come to endorse certain assumptions, including the idea that economic "growth" is essential to personal and national well-being and that contributing to society means spending money. As one of my students puts it:

Student Writing: Response Papers

Rather than being satisfied with what you have, "greed is good" thinking encourages people to never be content. Success is only measured by growth and increasing profits. If the U.S. gross domestic product does not grow by a certain rate each year it is seen as a bad sign. Americans must grow to accept what they already have rather than expecting infinite growth.

To keep this modern system of consumption in place and growing, we are surrounded by advertising and cultural discourses that encourage us to feel that we "need" more and more consumer goods. To be an accepted member of society, we have come to believe that we must own certain things, which we see as markers of our status and identity. It's important that students think critically about these messages urging us to consume, and recognize *how* these messages are acting on us. Here is an example of a student doing just this:

Student Writing: Response Papers

by Danny Clemens

I feel like marketers have done a fantastic job of selling us things under the premise that they will make us a better person.

Students should become aware that these modern forms of consumption have not always existed. The modes of making and consuming goods we've developed in our age of industrialization, globalization, and mass marketing may seem like the norm now, but this system – with its resource-intensive factory production, toxic by-products, outsourced sweatshop labor, disposability, and massive land-fills and incinerators for all the "trash" we create – is relatively new in human history. We are only starting to understand the impacts of such a toxic produc-tion chain, and of what William McDonough and Michael Braungart (2002) call a "cradle-to-grave" system in which we design products not for complete reuse but to simply be thrown "away." Here one of my students comments on this realization of how things have changed:

Student Writing: Response Papers

by Jacob Rosenborough

The simple fact that most people don't know how quickly we've made changes like these and how different we are from our not so far off past isn't very good. I fell [*sic*] like our lack of awareness of how the past used to be and how quickly we've changed makes us think that we're consuming and living as always, in a normal way that has always worked.

As our society consumes all of these mass-produced goods, we rarely take time to think closely about where the materials go when we dispose of them. Here a student points out our common avoidance of this subject:

Student Writing: Response Papers

by Jennie Williams

Trash is a topic people do not like to think about. Trash is generally gross for one, but it is also a source of guilt for our society. We throw things in our little garbage bins and then take it out to a larger trashcan usually outside, then that's the last we will ever see of our discarded waste. We don't like to imagine what happens next because it is unsettling.

Students should start to think about the enormous problems that exist with this system of sending valuable materials mixed with toxic chemicals to be buried in a landfill. They should also reflect on possible strategies for changing this consumer cycle, from recommendations for "living simply" suggested by authors like Jim Merkel (2003) to calls for redesigning the way we make goods altogether, as McDonough and Braungart (2002) propose.

Our modern consumer system has vast consequences globally. Those who aren't directly touched by the effects of the resource extraction, the pollution, the exploitation of workers, or the toxic disposal sites can too easily ignore the consequences of buying and throwing away so many consumer goods. Thomas Princen (2002) points out that the globalization of this production chain means we often don't experience local feedback that we can feel personally. The "distancing" of consumers from impacts allows privileged segments of society to blissfully continue buying. I hope the lessons in Unit 5 help throw open the veil of this distancing and encourage students both to become more informed and ethical consumers and to seek better ways to live that don't depend on making and acquiring goods the way we do today.

UNIT 5 READINGS AND REFERENCES

Durning, Alan Thein. 2006. "The Dubious Rewards of Consumption." In *The Earthscan Reader in Sustainable Consumption*, edited by Tim Jackson, 129–134. London: Earthscan.

Jordan, Chris. 2005. "Intolerable Beauty: Portraits of American Mass Consumption." *Chris Jordan Photographic Arts*. www.chrisjordan.com/gallery/intolerable/#cellphones2.

Jordan, Chris. 2011. "Midway: Message from the Gyre." *Chris Jordan Photographic Arts*. www.chrisjordan.com/gallery/midway/#CF000313%2018x24.

Manno, Jack. 2002. "Commoditization: Consumption Efficiency and an Economy of Care and Connection." In *Confronting Consumption*, edited by Thomas Princen, Michael F. Maniates, and Ken Conca, 1st ed., 67–99. Cambridge, MA: MIT Press.

Marshall, Andrew Gavin. 2012. "No Conspiracy Theory – A Small Group of Companies Have Enormous Power Over the World." *AlterNet*. October 31. www.alternet.org/world/no-conspiracy-theory-small-group-companies-have-enormous-power-over-world.

McDonough, William, and Michael Braungart. 2002. *Cradle to Cradle: Remaking the Way We Make Things*. 1st ed. New York: North Point Press.

McKibben, Bill. 2008. *Deep Economy: The Wealth of Communities and the Durable Future*. Later printing. New York: St. Martin's Griffin.

Merkel, Jim. 2003. *Radical Simplicity: Small Footprints on a Finite Earth*. Gabriola Island, BC: New Society Publishers.

Midway Film and Chris Jordan. 2012. *Midway Trailer*. Video recording. www.midwayfilm. com/index.html.

Orr, David. 2010. "The Carbon Connection." *Center for Ecoliteracy*. Accessed August 29. www.ecoliteracy.org/essays/carbon-connection.

Princen, Thomas. 2002. "Distancing: Consumption and the Severing of Feedback." In *Confronting Consumption*, edited by Thomas Princen, Michael F. Maniates, and Ken Conca, 1st ed., 103–131. Cambridge, MA: MIT Press.

Ryan, John C., and Alan Thein Durning. 1997. *Stuff: The Secret Lives of Everyday Things*. Seattle, WA: Northwest Environment Watch.

Story of Stuff Project. 2007. *The Story of Stuff*. Video recording. www.youtube.com/ watch?v=9GorqroigqM&feature=youtube_gdata_player.

Turner, Rita. 2010. "Discourses of Consumption in US-American Culture." *Sustainability* 2 (7): 2279–2301.

Wackernagel, Mathis, and Williams E. Rees. 1996. *Our Ecological Footprint: Reducing Human Impact on the Earth (New Catalyst Bioregional Series)*. Gabriola Island, BC: New Society Publishers.

Wright, Ronald. 2005. *A Short History of Progress*. New York: Da Capo Press.

Unit 6

ENVIRONMENTAL ATTITUDES AND BEHAVIORS IN U.S.-AMERICAN HISTORY

If we don't own the sweet air
and the bubbling water,
how can you buy it from us?
Each pine tree shining in the sun,
 . . . is holy in the thoughts
and memory of our people . . .
The fragrant flowers are our sisters,
the reindeer, the horse,
the great eagle our brothers . . .
So when the Great Chief in Washington
sends word that he wants to buy
our land, he asks a great deal of us.
The earth is not his brother
but his enemy
and when he has conquered it,
he moves on.
He cares nothing for the land.

−Chief Seattle, in Benton and Short (12)
Lesson 6.1

This unit explores how attitudes toward the environment have changed over time in the U.S. Students study opinions about the land and humankind's role within it from historical writings over the early history of the U.S. into the twentieth century. This includes analyzing attitudes expressed in Native American stories and speeches, documents written by early European colonists, writings of transcendentalist authors, and speeches and letters written by U.S. presidents. Students should be able to identify many distinct views of the land in these readings, and to think critically about how these views have influenced behaviors as well as which ones continue to be held today. In Lessons 6.2 and 6.3, students will apply their understanding of these different views of the land to analyses of modern environmental debates and of their own beliefs.

Some key questions that the lessons in this unit should raise for students:

- How have different groups of people perceived their relationship with the natural world over the history of the U.S.?
- What are the ramifications of holding different views of the land? What types of behaviors do our views of the land lead to?
- What views of the land do you think produce the most healthy relationships between people and the natural world? Why?
- Are the same views of the land held by groups in past history still held by people today? How do these views influence modern decision-making about environmental issues?
- What views of the land do you personally hold? Have those views changed during your life? Are they influenced by society?

LESSON 6.1: VIEWS OF THE LAND IN U.S.-AMERICAN HISTORY

About this Lesson

In this lesson students explore how different groups of people have historically viewed their relationship to the land in the United States, from Native American belief systems to colonists to conservationists and government policy-makers. In the lesson students read historical writings and articulate what attitude toward the land is manifested in each piece of writing.

Lesson Objectives

Students will:

- Learn about views of the land held by a range of groups throughout U.S. history.
- Analyze historical documents to determine the belief systems underlying what is said.
- Think critically about the implications of diverse belief systems about the land.
- Compose original analytical writing.

Lesson Activities at a Glance

1. Students read assigned texts that demonstrate different attitudes toward the natural world held in early to mid-U.S. history.
2. Students write response papers outside of class.
3. In class, students work in groups to analyze specific texts and identify what views of the land the authors operate upon.
4. Share analysis as a class and make a collective list of each view of the land seen in the readings.

Key Content Area Skills: reading and analyzing nonfiction prose, cultivating historical knowledge, writing analytical and reflective essays.

Texts Used in this Lesson

Texts for this lesson should be assigned in approximately this order:

Passages from *So Glorious a Landscape: Nature and the Environment in American History and Culture*, edited by Chris Magoc:

> "Acoma Pueblo Creation Myth," 20–22
> "Tewa Sky Looms," 23
> "A Hideous and Desolate Wilderness (1647)" by William Bradford, 24–26

Passages from *Environmental Discourse and Practice: A Reader*, edited by Lisa Benton and John Short:

> "How Can One Sell the Air" by Chief Seattle, 12
> "A Certaine Indian (1621)" by William Bradford, 19–20
> "Before They Got Thick" by Percy Bigmouth, 20–21
> "Changes in the Land: Indians, Colonists, and the Ecology of New England" by William Cronon, 37–44

Barbara Sproul, *Primal Myths*:

> "Making the World," 245–248
> "The Making of Men and Horses," 252–253
> "How the World Was Made," 253–255
> "Creation of the Earth," 255–257
> "The Earth Is Set Up," 258–260
> "The Way of the Indian," 260–262

Virginia Armstrong, *I Have Spoken*, 1–23 and 34–39

Passages from *So Glorious a Landscape: Nature and the Environment in American History and Culture*, edited by Chris Magoc:

> "The Untransacted Destiny of the American People (1846)" by William Gilpin, 43–44
> "Americans Spread All Over California (1846)" from the Monterey Californian, 45
> "Social and Environmental Degradation in the California Gold Country (1890)" by Joaquin Miller, 46–49
> "Where I Lived and What I Lived For (1854)" by Henry David Thoreau, 74–79
> "My First Summer in the Sierra (1868)" by John Muir, 80–83
> "The Land of Little Rain (1903)" by Mary Austin, 92–95
> "The Destructiveness of Man (1864)" by George Perkins Marsh, 136–139

Optional: Andrea Wulf, *Founding Gardeners*, 3–34

Optional: John Sears, *Sacred Places*, 3–30

Passages from *Environmental Discourse and Practice: A Reader*, edited by Lisa Benton and John Short:

> "Moving West (1797)" by Daniel Boone, 59–60
> "The 1785 Ordnance," 60–62
> "The Oregon Trail (1849)" by Francis Parkman, Jr., 62–63
> "Letters Home (1863–1865)" by Gro Svendsen, 64–66
> "The Significance of the Frontier in American History (1894)" by Frederick Jackson Turner, 75–77
> "Essay on American Scenery (1835)" by Thomas Cole, 87–90

"National Park Legislation (1864)," 98
"National Park Legislation (1872)," 98–99
"A Voice for Wilderness (1901)" by John Muir, 102–104
"National Park Legislation (1916)," 104–105
"Conservation, Protection, Reclamation, and Irrigation (1901)" by Theodore Roosevelt, 110–113
"Theodore Roosevelt and Conservation" by H. W. Brands, 113–116
"The Birth of Conservation" by Gifford Pinchot, 116–118

Optional: Robert Gottlieb, *Forcing the Spring*, 47–70

Optional: John Steinbeck, *The Grapes of Wrath*, 5–42 (or entire novel)

Note: Publication information for these readings is listed at the end of the unit. Sample passages are included in the full List of Readings at the end of the book.

My Procedure

I start this lesson by asking students to think about how we in the U.S. view our relationship with the natural world, and how we have viewed it over the course of history. I ask them to think about what "the land" and the nonhuman world means to them, and why they feel this way. We discuss students' initial sense of these questions in class. I then have students read the assigned texts for this lesson. These can be split over multiple class sessions or weeks, but the readings at the top of the list should be assigned first. The readings for this lesson are arranged chronologically, starting with views expressed by early Native Americans and early colonists, then moving into the perspectives of American revolutionaries, "pioneers" moving west, and conservationists. It's best to teach this lesson over multiple class sessions; assign the readings from the top of the list through and including Virginia Armstrong first, followed by the readings listed below those. (Those first readings may still be split over more than one class session, but should be assigned before the readings that are listed below Virginia Armstrong.) Any study of changing views of the land in U.S. culture could certainly continue beyond the stage of history covered in these readings, but for the purposes of this lesson, looking at early history from the 1600s to the early 1900s gives students a sense of the roots of many modern views of the land, and also provides a clear picture of the contrasting views held by different groups, including common Native American perspectives and how those perspectives differed from colonists'. Note that many of these passages are excerpts of historical texts, and are therefore written with different spelling and grammar conventions. Some of them can be challenging to parse, so students may need additional support, or you may choose to narrow the readings. I do recommend keeping in at least a couple of representative texts from each time period, so students can be exposed to these diverse perspectives.

Students read these texts outside of class and write response papers. In their response papers, they quickly begin to notice clear differences between views of the land held by different groups in history. The differences between Native American and colonial views are the first contrasts that students encounter in the readings and the first they analyze. Here's an example of a student responding to these readings:

Student Writing: Response Papers

by Ashley Sweet

It is apparent that Native Americans revered their land and viewed it as a continuation of them. The idea of claiming ownership was unfathomable, and their distaste for the settlers desire to purchase the land was clear and certainly a contributor to the discord between them. Referring back to I Have Spoken, Half-King, in a speech to the French Commandant states, "the land belongs to neither one or the other, but the Great Being above allowed it to be a place of residence for us" (Armstrong, p. 17). . . . Where the Native Americans worship the land, respect it and see the value in all it holds and offers, in its constant ability to grow and be reborn and provide over and over the things man needs, the Colonials saw the land as an investment. They desired to build upon the land, to settle it and inhabit it, to stake claim on the ground and "enhance" it's [*sic*] value with man-made things, buildings and such. In I Have Spoken, George Thomas speaks to the Native Americans and says, "It is very true, that Lands are of late become more valuable; but what raises their Value? Is it not entirely owing to the Industry and Labour used by the white People, in their Cultivation and Improvement? Had not they come amongst you, these Lands would have been of no Use to you, any further than to maintain every Thing; but you know very well, that they cost a great deal of Money; and the Value of Land is no more that it is worth in Money" (Armstrong, p. 15). Thomas' thoughts on the land, that their value lies only in its monetary worth, contradict the very foundation of Native American culture.

When students come to class after completing the readings, I have them get into small groups, and I assign each group one of the passages from the readings. I ask each group to try to articulate how they think the author of the piece, or the group of people discussed in the piece, would describe the purpose of the land and how humans "should" interact with it. What would they say the natural world means to them?

I move between groups to help students through this discussion. I tell them to find evidence in the text that reveals what view of the land is motivating what the author or subject of the piece says. It's important to help students think critically about what is being expressed, even in cases where it may sound "normal" to them and may not seem far removed from their own opinions.

Once students have talked through their assigned passages, I have each group share their conclusions with the whole class. As we do this, we make a list together of "land as . . . " statements – what the author or subject of each reading sees the land as. For example, students may find that in some of the Native American passages the authors talk about the land as a family member or as a living body.

In early colonists' writings, the land is often spoken of as a dark and dangerous place or as a commodity to be catalogued and collected for sale. As the readings move forward in history, students may find transcendentalists speaking of the land as a place of worship, or conservationists talking about the land as a resource for human well-being that must be carefully conserved.

We go through each of these views, listing them together. I use a PowerPoint slide or Word document that I project onscreen in the classroom, and I type the list live as students contribute to it. I sometimes suggest items for the list myself if there are views expressed in the readings that students haven't raised, and I sometimes re-word students' suggestions if I think they're going in the right direction but can't quite find the words for what they are trying to express. However, most of the items on the list I type up exactly as students articulate them.

When I teach this lesson I assign the readings listed above over several class sessions. My students and I return to our "land as . . . " list with every new set of readings, adding new entries to the list that cover the views of groups at different points in history.

In the box below I've reprinted just some items from the long list created with my students during one of my semesters of teaching this lesson, quoted directly from the list I typed during class.

The Land As . . .

Native Americans' Perspectives

- land as a body
- land as united with humanity, as one
- land as a balance
- earth as mother
- earth as family
- earth as gift
- land as reflection of self

Colonists' Perspectives

- earth as limitless
- earth as source of profit
- land as "wasteland," "wilderness"
- land as commodity
- land as danger
- land as unholy, evil
- land as financial security, assets for future
- land as object to control, have dominion over
- land as having no inherent value, needing value brought to it

1700's and 1800's Perspectives

- land as having aesthetic value, "grand," "beautiful"
- land as something to improve, to manage

(continued)

(continued)

- land as independence, political and financial freedom
- land as entertainment, therapy
- land as territory to fight for/over
- land as purpose, occupation
- land as untapped potential
- land as having spiritual significance
- land as source of emotional, psychological nourishment
- land as escape
- land as source of quiet, peace
- land as adventure
- land as unknown, mystery
- land as vacant
- land as untamed
- land as political power, political domination
- land as national identity
- land as luxury, indulgence
- land as memorial, history

As this list demonstrates, the texts assigned for this lesson represent many different views of the land held by different groups of people over long periods of U.S. history. The goal here is not to identify one monolithic view held by a certain group of people – there is no one view held by Native Americans, or by European colonists. However, there are types of views that are common among certain groups of people, and are noticeably different from other groups. This is worth identifying. It's also important to think about how these views have played out over time, and what the results of holding certain views of the land has been for people, other beings, and ecosystems. As we list and discuss these views of the land, we also talk about their implications, and what actions they have motivated over time.

Products and Assessment

Students' participation in class discussion and their completion of written response papers are the means for evaluating their achievement of the objectives for this lesson. In their class discussion, as they seek to describe the views of the land represented in the passages they are discussing, they should be justifying their conclusions with evidence from the text and demonstrating careful reading of the material.

I use the following grading criteria to evaluate student work for this lesson. I have, in different classes, formatted these criteria as a grading rubric or listed them in my course syllabus. How to format and apply the grading criteria will depend on individual context, school requirements, and teacher preference.

Grading Criteria

- Student demonstrates critical thinking about how groups of people throughout history have thought about the land, ecosystems, and other beings.
- Student demonstrates critical analysis of text to draw conclusions about views of the land expressed by authors and subjects of assigned readings.
- Student demonstrates thoughtful reflection about the implications of different views of the land.
- Work demonstrates thorough reading and comprehension of assigned course texts.
- Work demonstrates personal reflection, critical thought, and insight into course texts.
- Arguments are clear, well-developed, and documented with evidence from texts.
- Work demonstrates critical analysis of course topics, questions, and subject-matter.
- Style, usage, format, grammar, imagery, and presentation support meaning and are intentional, creative, and original.
- Work meets all requirements of the assignment and is utilized to facilitate development of personal understanding.

LESSON 6.2: VIEWS OF THE LAND IN MODERN ENVIRONMENTAL DEBATES

About this Lesson

This lesson builds directly from Lesson 6.1. Having compiled a list of different views of the land that have been held by groups of people at points throughout U.S. history, students begin this lesson by selecting modern environmental issues, either local or national, that they want to analyze. Students then collect examples of comments made about their chosen issue from all "sides," and try to identify what views of the land are motivating each party to take the stand that they are taking on this issue. This gives students an opportunity to see the way that historical views of the land are often still active in modern ideologies and that differing belief systems will lead people to draw different conclusions about how humans can and should interact with the land.

Lesson Objectives

Students will:

- Research the opinions and arguments of diverse stakeholders in the context of a specific environmental issue or debate.
- Critically analyze statements made in public media and draw conclusions about views of the land motivating those statements.
- Create and present a multi-media presentation outlining an environmental issue, the opinions of stakeholders around this issue, and an analysis of the views that motivate those opinions.

Lesson Activities at a Glance

1. Students form groups and select an environmental issue that is currently under debate to research.
2. Students find statements made by individuals participating in the debate around this issue and analyze these statements.
3. Students present their analysis to the class.

Key Content Area Skills: critical analysis of language and media, critical analysis of cultural belief systems, creating multi-media presentations.

Texts Used in this Lesson

This lesson has no new assigned texts, but builds on those discussed in Lesson 6.1.

My Procedure

This lesson follows on from Lesson 6.1 in which students identify a number of different views of the land that groups of people have held at various points in U.S. history. Having discussed these different views, I start Lesson 6.2 by having students get into groups. I have each group select an environmental issue currently under debate that they would like to analyze. This can be a local, regional, or national issue. I have had groups select everything from a debate over whether to build a local road that would involve cutting down a section of forest, to factory farm runoff polluting the Chesapeake Bay, to hydraulic fracturing and oil drilling.

I give each group the assignment below; they are to explore views of the land held by parties on both sides of the environmental issue they've selected. Groups must find statements made by participants in the debate around their chosen issue. In most cases, this primarily means finding statements that are "pro" and "con" – for example, those who are arguing *for* building the road and those who are arguing *against* it. Students can draw from statements made by these groups in any form of media, such as newspaper articles or television interviews. I have them collect statements from each "side," and then analyze what view of the land they see manifested in the statements made by each of these stakeholders. Here is the assignment:

Assignment: Views of the Land in Modern Environmental Debates

At the beginning of the unit you'll be paired with a partner to conduct research on a current environmental debate in the U.S. Together you will select one topic, either local or national, that you will research.

Your research should seek to answer the following questions:

- What decision or question is at stake?
- What is the controversy about this decision?
- Who is participating in the debate?
- What arguments are being made?
- What views of the land are held by each party, and how are these views shaping their opinions?
- What evidence do you have of these views at work? How are each party's views of the land manifesting in their language and actions?

You and your partner will then create a short presentation to share your research with the class. This presentation should include quotes and clips from print, internet, and television news sources that demonstrate what views of the land each participant in the debate holds. Your presentation will take the form of a PowerPoint slideshow and must include a written narrative, either within the "notes" field of the slideshow file or as a separate word-processing file to be submitted to me. Your presentation must cite at least *five* relevant sources and must include a full list of references.

Students present their research and analysis to the class with a PowerPoint or similar presentation. In their presentations they outline the issue, share examples of the statements made by each stakeholder, and then share their analysis of what views of the land they feel each stakeholder is operating from.

Here's an excerpt from one presentation that explored a debate over building a new tar sands oil pipeline. In the presentation students compared views of the land held by the company that wanted to build the pipeline, versus views held by residents of the region the pipeline would run through, including indigenous populations, who did not want the pipeline built. Students used news clips and quotes from both sides of the debate to determine what views were held by these groups. Here are the views outlined in the presentation:

Student Presentation: Views of the Land in Modern Environmental Debates

by Danny Clemens

[Oil Company's] Views

- Land as a commodity
- Land as limitless
- Land as needing to be harnessed
- Land as dumping ground
- Land as needing purpose

Residents' and [Environmental Group's] Views

- Land as needing to be protected
- Land as fallible
- Land as defining culture/identity
- Land as a victim

In these presentations students should be thinking about what attitudes are motivating people to believe that the land should be treated certain ways. By drawing on the history they learned in Lesson 6.1, students can gain much greater perspective on how modern environmental debates are rooted in long legacies of beliefs about humankind's relationship with the natural world.

Products and Assessment

The primary product that comes from this lesson is the presentations in which students share their research and analysis. I require students to include a written document containing their "script" for what they will say during their presentation, as well as a file containing the presentation itself. The presentation should also include references for all sources.

I use the following grading criteria to evaluate student work for this lesson. I have, in different classes, formatted these criteria as a grading rubric or listed them in my course syllabus. How to format and apply the grading criteria will depend on individual context, school requirements, and teacher preference.

Grading Criteria

- Student demonstrates critical analysis of language and ideology.
- Student conducts valid research into an environmental issue, citing all sources.
- Student effectively analyzes beliefs held by stakeholders in her/his selected environmental issue, using evidence from statements made in the media to justify conclusions.
- Student demonstrates thoughtful reflection on the implications of different views of the land.
- Student demonstrates understanding of common views of the land that have been held in U.S. history past and present.
- Work demonstrates thorough reading and comprehension of assigned course texts.
- Work demonstrates personal reflection, critical thought, and insight into course texts.
- Arguments are clear, well-developed, and documented with evidence from texts.
- Work demonstrates critical analysis of course topics, questions, and subject-matter.
- Style, usage, format, grammar, imagery, and presentation support meaning and are intentional, creative, and original.
- Work meets all requirements of the assignment and is utilized to facilitate development of personal understanding.

LESSON 6.3: DIGITAL STORY

Personal History of Views of the Land

About this Lesson

This lesson provides a way for students to relate the historical views of the land they learned about in Lesson 6.1 to their own lives. In this lesson students create digital stories reflecting on different attitudes toward the land that they have held throughout their lives, and discussing any significant events or places that have influenced their thinking on the subject. Students then present these digital stories in class.

Lesson Objectives

Students will:

- Reflect on personal beliefs about the land and formative experiences that have shaped those beliefs.
- Connect their personal beliefs and experiences to knowledge of historical beliefs about humankind's relationship with the natural world.
- Think critically about the implications of different views of the land.
- Plan, create, and present a digital story, containing photographs, music, and a narrative, reflecting on their views of the land throughout their lives.

Lesson Activities at a Glance

1. Students review the diverse views of the land discussed in Lesson 6.1.
2. Students create digital stories.
3. Students present digital stories in class.

Key Content Area Skills: creating multi-media presentations, thinking critically about history and ideology, using media as a tool for personal reflection.

Texts Used in this Lesson

This lesson has no new assigned texts, but builds on those assigned in Lesson 6.1.

My Procedure

To start this lesson, I review with students the views of the natural world we've encountered in this unit so far. We discuss the list of "land as . . ." views compiled during Lesson 6.1, and summarize some of the views of the land students identified in their presentations in Lesson 6.2. I then ask students to think about the views they themselves have held about the natural world over the course of their lives. Then I introduce the following assignment:

Assignment: Digital Story of How You View the Land

Create a digital story in which you reflect on the places that have shaped your life and how you have thought about these places. Describe your relationship to the natural world, in light of the perspectives we have explored in class. How have you thought about the natural world in the past, and how do you think of it now? Which attitudes, held by which cultural groups we've encountered, are most similar to your own? Having studied the history of these cultural relationships with the land, how do you now interpret your own relationship to the land within this historical context? What do you think of the current state of our relationships with the land in the U.S.?

For your digital story, you must use original photographs and/or video footage of places that have been important parts of your life. You may include photographs of places you've spent time, important trips you've taken, and your home and neighborhood. Use creativity and artistry to convey what it feels like to be in these spaces and what these places mean to you. Your digital story must incorporate insights from the class, and must directly discuss at least *four* course texts. Discuss the views and insights expressed by these texts, and relate those views to your own. Use these texts to answer the above questions and to reflect on your own relationships to the land.

Submit a written narrative along with your digital story, and make sure you properly cite all sources.

In the digital stories they create for this assignment, students typically discuss several key places in the natural world that they have spent time in during their lives, tracing how each of these places influenced their thinking and how their attitudes toward the natural world have grown and changed over time. For this assignment I require students to reference some of the assigned readings from Lesson 6.1, and demonstrate that they are relating those readings to their own experiences.

Here's an example from the narrative accompanying one of my students' digital stories:

Student Writing: Digital Story Narrative

by Ashley Sweet

I was most moved by writings such as those of John Muir, which attempted to capture the grandeur and beauty of a natural landscape. He often mentioned how any description he could offer, no matter how glorious, would fall short of expressing what he saw. I feel that way frequently. Even in the simplest places, I'm often in awe of the natural world and the way it makes me feel whole and peaceful. When I am overwhelmed, I escape to the recesses of the world outside. Unlike most people I know, I don't need to leave home for this. . . . I just sit or lie on the ground and feel the energy of the earth beneath me.

This assignment offers students an opportunity to synthesize the ideas they've explored in this unit and look closely at their own relationship with the natural world and what has influenced it.

Note that there are some similarities between this assignment and the digital story assignment in Lesson 3.3. I have never assigned both of these digital stories in the same class – I've used this assignment in a more advanced class on the history of attitudes toward the land in the U.S., and I have used the digital story assignment in Lesson 3.3 in an introductory class on environmental sustainability. Feel free to assign both in the same class, but since there is potential for overlap, you may want to pick one or the other if you're teaching both lessons.

Products and Assessment

The product created in this lesson is students' digital stories. These should be personal and thoughtful, and should also make direct use of insights from the assigned readings from earlier in this unit.

I use the following grading criteria to evaluate student work for this lesson. I have, in different classes, formatted these criteria as a grading rubric or listed them in my course syllabus. How to format and apply the grading criteria will depend on individual context, school requirements, and teacher preference.

Grading Criteria

- Student demonstrates critical self-reflection about personal beliefs.
- Student demonstrates thoughtful reflection on the implications of different views of the land.
- Student demonstrates understanding of common views of the land that have been held in U.S. history past and present.
- Student relates insights from course texts to personal experience.
- Work demonstrates thorough reading and comprehension of assigned course texts.
- Work demonstrates personal reflection, critical thought, and insight into course texts.
- Arguments are clear, well-developed, and documented with evidence from texts.
- Work demonstrates critical analysis of course topics, questions, and subject-matter.
- Style, usage, format, grammar, imagery, and presentation support meaning and are intentional, creative, and original.
- Work meets all requirements of the assignment and is utilized to facilitate development of personal understanding.

EFFECT OF THE LESSONS: CRITICAL HISTORICAL AND CULTURAL PERSPECTIVES

It's easy to feel like the beliefs we are exposed to every day in our culture are held by everyone, and always have been. Thinking of the land as a commodity, for example, that has no value beyond its potential to make money for the person who "owns" it – this may seem like a normal belief that people have always held. But this is not historically the case – beliefs such as this come from certain traditions, and are not the only ways to view the world. This is why learning about the beliefs systems of other time periods and other cultural groups is so essential. It's important to expose ourselves to other beliefs and ideologies, past and present, to help us reflect more critically on our own beliefs and expand our awareness of alternative approaches.

In this unit students see the starting place of belief systems that have been extremely influential in shaping U.S.-American behaviors and attitudes toward the natural world, such as the commonly held view among colonists that the natural world was useful purely insofar as it could produce monetary gain. They're also presented with views of the natural world that contrast sharply with dominant perspectives in the U.S., like views of Native American authors that conceptualize the natural world as a family member, divine gift, or living being. Considering differing views such as these can inform our critical awareness about our own culture and fuel creative explorations of possible alternative conceptions.

Discussing social crises, philosopher Alasdair MacIntyre notes "The ability to respond adequately to this kind of cultural need depends of course on whether those summoned possess intellectual and moral resources that transcend the immediate crisis, which enable them to say to the culture what the culture cannot say to itself" (qtd. in Smith 2001, 14). I hope this unit helps provide students with just these sorts of resources, through development of informed cultural perspective. As they engage with these lessons, I hope students will gain the insights to view modern society from a new angle and "say to our culture what it cannot say to itself."

As they work on these lessons, look for students to build comparative knowledge of different views, and use this comparative understanding to gain new perspective on our relationship with the natural world today. Here's a great example of this process, as a student applies understandings of colonial and Native American perspectives to think critically about modern beliefs:

Student Writing: Response Papers

by Ashley Sweet

The white settler views of the land, it's worth in money, got me thinking about the way we live today. What land we own and how much are key reflections of our status in normal society. Americans today have lost a lot of the awe of the land itself in a natural state and see only the value land has in its ability to be built upon or the monetary value of the natural resources (oil, diamonds, etc.) it contains. It's interesting to consider the shift in the perception of American soil and its value from the Native American perspective to the "white" or modern perspective today.

As students explore historical perspectives, they should also start to recognize negative consequences that result from certain views of the land. Here's one student discussing the consequences of believing one can "own" the land:

Student Writing: Response Papers

by Will Fejes

[T]he idea of land ownership can in many cases lead to negative effects on the land. I think that this is caused because of the way our society is built is that of if you own something it is all yours to do whatever you want with it.

In addition to getting students thinking more critically about common views of the land they may already see around them in modern culture, Lesson 6.1 may also introduce students to different perspectives on the land that they hadn't considered before. Here's one student discussing perspectives beyond the dominant commodifying view:

Student Writing: Response Papers

by Malarie Novotny

I think that the land also offers a way to be creative, which is different from what we usually discuss about its resources and everything. I think it's interesting the land has become a source for creativity, passion, love, and even employment.

This unit ends with the digital story assignment in Lesson 6.3 in order to encourage students to apply their newly acquired cultural perspective in order to reflect on their own views of the land and how these have been influenced by society. In the following example, a student reflects on the ways that even the structure of our daily lives can lead to disconnection and change our relationship with the natural world:

Student Writing: Response Papers

by Ashley Sweet

Living in a country where, perhaps regrettably, most things are provided for me, I, through the progress of my ancestors, have lost any inherent ability to understand the sounds of nature and the animals in it.

I hope students come away from this unit thinking – informed by history and new cultural knowledge – about why society holds the views of the land it does, what impact those views have on behavior, why their own relationship to the natural world is what it is, how their culture has influenced that relationship, and what other relationships are possible. Such thinking means putting cultural perspective to good use, and learning to "say to our culture what it cannot say to itself."

UNIT 6 READINGS AND REFERENCES

Armstrong, Virginia I. 1971. *I Have Spoken: American History Through The Voices Of The Indians.* Athens, OH: Swallow Press.

Austin, Mary. 2001. "The Land of Little Rain (1903)." In *So Glorious a Landscape: Nature and the Environment in American History and Culture,* edited by Chris J. Magoc, 92–95. Wilmington, DE: Rowman & Littlefield.

Benton, Lisa M., and John Rennie Short, eds. 2000a. "The 1785 Ordnance." In *Environmental Discourse and Practice: A Reader,* 60–62. Malden, MA: Wiley-Blackwell.

Benton, Lisa M., and John Rennie Short, eds. 2000b. "National Park Legislation (1864)." In *Environmental Discourse and Practice: A Reader,* 98. Malden, MA: Wiley-Blackwell.

Benton, Lisa M., and John Rennie Short, eds. 2000c. "National Park Legislation (1872)." In *Environmental Discourse and Practice: A Reader,* 98–99. Malden, MA: Wiley-Blackwell.

Benton, Lisa M., and John Rennie Short, eds. 2000d. "National Park Legislation (1916)." In *Environmental Discourse and Practice: A Reader,* 104–105. Malden, MA: Wiley-Blackwell.

Bigmouth, Percy. 2000. "Before They Got Thick." In *Environmental Discourse and Practice: A Reader,* edited by Lisa M. Benton and John Rennie Short, 20–21. Malden, MA: Wiley-Blackwell.

Boone, Daniel. 2000. "Moving West (1797)." In *Environmental Discourse and Practice: A Reader,* edited by Lisa M. Benton and John Rennie Short, 59–62. Malden, MA: Wiley-Blackwell.

Bradford, William. 2000. "A Certaine Indian (1621)." In *Environmental Discourse and Practice: A Reader,* edited by Lisa M. Benton and John Rennie Short, 19–20. Malden, MA: Wiley-Blackwell.

Bradford, William. 2001. "A Hideous and Desolate Wilderness (1647)." In *So Glorious a Landscape: Nature and the Environment in American History and Culture,* edited by Chris J. Magoc, 24–26. Wilmington, DE: Rowman & Littlefield.

Brands, H. W. 2000. "Theodore Roosevelt and Conservation." In *Environmental Discourse and Practice: A Reader*, edited by Lisa M. Benton and John Rennie Short, 113–116. Malden, MA: Wiley-Blackwell.

Chief Seattle. [1855] 2000. "How Can One Sell the Air?: A Manifesto for the Earth." In *Environmental Discourse and Practice: A Reader*, edited by Lisa M. Benton and John Rennie Short, 12–13. Malden, MA: Wiley-Blackwell.

Cole, Thomas. 2000. "Essay on American Scenery (1835)." In *Environmental Discourse and Practice: A Reader*, edited by Lisa M. Benton and John Rennie Short, 87–90. Malden, MA: Wiley-Blackwell.

Cronon, William. 2000. "Changes in the Land: Indians, Colonists, and the Ecology of New England." In *Environmental Discourse and Practice: A Reader*, edited by Lisa M. Benton and John Rennie Short, 37–44. Malden, MA: Wiley-Blackwell.

Gilpin, William. 2001. "The Untransacted Destiny of the American People (1846)." In *So Glorious a Landscape: Nature and the Environment in American History and Culture*, edited by Chris J. Magoc, 43–44. Wilmington, DE: Rowman & Littlefield.

Gottlieb, Robert. 2005. *Forcing the Spring: The Transformation of the American Environmental Movement*. Revised ed. Washington, DC: Island Press.

Magoc, Chris J., ed. 2001a. "Acoma Pueblo Creation Myth." In *So Glorious a Landscape: Nature and the Environment in American History and Culture*, 20–22. Wilmington, DE: Rowman & Littlefield.

Magoc, Chris J., ed. 2001b. "Tewa Sky Looms." In *So Glorious a Landscape: Nature and the Environment in American History and Culture*, 23. Wilmington, DE: Rowman & Littlefield.

Marsh, George Perkins. 2001. "The Destructiveness of Man (1864)." In *So Glorious a Landscape: Nature and the Environment in American History and Culture*, edited by Chris J. Magoc, 136–139. Wilmington, DE: Rowman & Littlefield.

Miller, Joaquin. 2001. "Social and Environmental Degradation in the California Gold Country (1890)." In *So Glorious a Landscape: Nature and the Environment in American History and Culture*, edited by Chris J. Magoc, 46–49. Wilmington, DE: Rowman & Littlefield.

Monterey Californian. 2001. "Americans Spread All Over California (1846)." In *So Glorious a Landscape: Nature and the Environment in American History and Culture*, edited by Chris J. Magoc, 45. Wilmington, DE: Rowman & Littlefield.

Muir, John. 2000. "A Voice for Wilderness (1901)." In *Environmental Discourse and Practice: A Reader*, edited by Lisa M. Benton and John Rennie Short, 102–104. Malden, MA: Wiley-Blackwell.

Muir, John. 2001. "My First Summer in the Sierra (1868)." In *So Glorious a Landscape: Nature and the Environment in American History and Culture*, edited by Chris J. Magoc, 80–83. Wilmington, DE: Rowman & Littlefield.

Parkman, Francis, Jr. 2000. "The Oregon Trail (1849)." In *Environmental Discourse and Practice: A Reader*, edited by Lisa M. Benton and John Rennie Short, 62–63. Malden, MA: Wiley-Blackwell.

Pinchot, Gifford. 2000. "The Birth of Conservation." In *Environmental Discourse an Practice: A Reader*, edited by Lisa M. Benton and John Rennie Short, 116–118. Malden, MA: Wiley-Blackwell.

Roosevelt, Theodore. 2000. "Conservation, Protection, Reclamation, and Irrigation (1901)." In *Environmental Discourse and Practice: A Reader*, edited by Lisa M. Benton and John Rennie Short. Malden, MA: Wiley-Blackwell.

Sears, John F. 1999. *Sacred Places: American Tourist Attractions in the Nineteenth Century*. Reprint ed. Amherst: University of Massachusetts Press.

Smith, Mick. 2001. *An Ethics of Place: Radical Ecology, Postmodernity, and Social Theory.* Albany, NY: State University of New York Press.

Sproul, Barbara C. 1991. *Primal Myths: Creation Myths Around the World.* 1st HarperCollins ed. San Francisco: HarperSanFrancisco.

Steinbeck, John. 1939. *The Grapes of Wrath.* New York: Viking Press.

Svendsen, Gro. 2000. "Letters Home (1863–1865)." In *Environmental Discourse and Practice: A Reader,* edited by Lisa M. Benton and John Rennie Short, 64–66. Malden, MA: Wiley-Blackwell.

Thoreau, Henry David. 2001. "Where I Lived and What I Lived For (1854)." In *So Glorious a Landscape: Nature and the Environment in American History and Culture,* edited by Chris J. Magoc, 74–79. Wilmington, DE: Rowman & Littlefield.

Turner, Frederick Jackson. 2000. "The Significance of the Frontier in American History (1894)." In *Environmental Discourse and Practice: A Reader,* edited by Lisa M. Benton and John Rennie Short, 75–77. Malden, MA: Wiley-Blackwell.

Wulf, Andrea. 2012. *Founding Gardeners: The Revolutionary Generation, Nature, and the Shaping of the American Nation.* Reprint ed. New York: Vintage.

Unit 7

ETHICS AND ENVIRONMENTAL JUSTICE

In the U.S. and elsewhere, only humans possess legal rights. People seeking legal redress for wrongs committed against nonhuman animals must sue in their own behalf . . . Legally, nonhuman animals are human property. In the U.S. a free-living deer or trout is public property. A cat or dog living with a human family is personal property. A cow commercially exploited for her milk is business property. . . . By defining nonhuman animals as property, the law sanctions their enslavement and murder. . . . Under the law, "persons" are rights-holders whereas "animals" are not. (Legally, animal always excludes humans. After all, we never would consent to be "treated like animals.") The boundary for rights belongs between beings and things, not between human and non-human animals. . . . A chicken or a honeybee needs and deserves legal rights. Equitable laws would redefine *person* and *individual* to include nonhuman animals or replace those terms with *animal* or *sentient being.*

—*Joan Dunayer (170–171)*
Lesson 7.1

This unit includes lessons that tackle questions of environmental ethics, animal rights, and environmental justice. Assignments accompanying these lessons ask students to compare the ethical arguments of different authors, reflect on the motivations behind moral and legal decision-making, and extrapolate how society would need to change in order to align with different ethical ideals. Lessons discuss how nonhuman animals should ethically be treated by humans, compare the rights of human and nonhuman groups in the U.S. and other countries, offer examples of environmental racism, and discuss the impact of globalization.

Some key questions that the lessons in this unit should raise for students:

- How *should* nonhuman beings be treated? What sorts of treatment are unethical?
- Do all living beings deserve legal rights? Are there types of protection from abuse that all living beings should have?
- How does a society decide who deserves rights and who doesn't? How and why does this change over time?
- Do humans have the right to use and damage the natural world in the ways we are?
- What people suffer most directly from pollution and environmental degradation? Why?
- What can be done to correct injustices against people and nonhuman beings?

LESSON 7.1: ENVIRONMENTAL ETHICS

Thinking about Right and Wrong

About this Lesson

In this lesson students read arguments exploring different ethical stances toward nonhuman beings and the natural world. They compare the arguments of each author, and they consider how ethical norms are formed in society and how the circle of ethical consideration has traditionally been delineated or how it gets expanded. Then they consider what changes would have to be made in society in order to achieve the type of ethical behavior each author argues for.

Lesson Objectives

Students will:

- Learn about ethical arguments regarding the treatment of nonhuman animals and the land.
- Analyze the implications of differing ethical arguments.
- Apply ethical arguments to draw conclusions about how nonhuman animals and the land should be treated.
- Reflect on personal ethical stances and behaviors.
- Compose original analytical writing.

Lesson Activities at a Glance

1. Students read assigned texts on ethical treatment of nonhumans and the land.
2. Students write response papers outside of class.
3. In class, students work in groups to compare arguments made by each author.
4. Students share their analyses with the class.

Key Content Area Skills: reading and analyzing nonfiction prose, understanding and applying theories of ethics, writing analytical and reflective essays.

Texts Used in this Lesson

Peter Singer, *Animal Liberation*, 1–23

Joan Dunayer, *Animal Equality: Language and Liberation*, 169–177

Aldo Leopold, *A Sand County Almanac:* "The Land Ethic," 237–246

Christopher Stone, *Should Trees Have Standing*, 1–31

Federico García Lorca, "New York," in *News of the Universe: Poems of Twofold Consciousness*, 110–112

Kevin Bowen, "Gelatin Factory," in *Poetry Like Bread*, 68–69

Optional: Anna Peterson, *Everyday Ethics and Social Change*, 1–25

Note: Publication information for these readings is listed at the end of the unit. Sample passages are included in the full List of Readings at the end of the book.

My Procedure

This lesson begins with students reading the assigned texts and writing response papers outside of class. The readings for this lesson include arguments for changing our treatment of nonhuman animals, plants, and the land, and authors draw from theories of ethics to discuss why many common practices – such as factory farming, animal testing, vivisection, animals in circuses, and the like – cannot be justified on ethical grounds.

There's a lot to process in these readings. Questions of "right and wrong" can be tricky to discuss, but they're an essential topic we can't shy away from if we are to reduce suffering and increase the possibility of living just and sustainable lives.

The readings in this lesson build on topics discussed in Lesson 2.3 and 2.4 – it's ideal if students have already participated in those lessons, although it is not essential. Authors of some of the assigned readings for this lesson reiterate some points from those lessons about the history of discrimination against human and non-human groups. Authors like Peter Singer point out that arguments were made against giving rights to women, African Americans, and other human groups in the past, and that similar reasoning makes it unethical to deny nonhuman animals due consideration. Joan Dunayer points this out as well, stressing that nonhuman animals are deemed property in our legal system and in our patterns of thought, and that this unjust categorization denies them their own rights and agency. Students often discuss these points in their response papers, and are quick to notice the negative ramifications. Here's an example from an anonymous student:

Student Writing: Response Papers

Throughout history, most judges and government officials have sided with the idea that only humans have legal rights. There is a ton of red tape and bureaucratic hoops to jump through to sue for damages of a nonhuman's pain and suffering. By calling animals "property" we have placed a hierarchal system in place. People are deemed superior to non-human beings because we "hold them" in our "possession" as "private property". Like any other progressive movement in history we need to again, redefine what encompasses an "individual" and "person". Dunayer fittingly addresses the abolishment of slavery, civil rights movement, and women's rights movement as continuing changes of the definition of "persons". So, does it not make sense that animal equality is the next step?

Students often question and analyze the reasons for exploitation of nonhuman animals and discuss whether and how they think such treatment could change. Here a student refers back to the metaphors for the natural world discussed in Lesson 2.2:

Student Writing: Response Papers

by Jacob Rosenborough

The reason behind our lacking's [sic] isn't just due to laws, but our mind sets. As I read the pieces I kept flashing back to when we discussed Worldviews and human/nature relationships. If America didn't operate under the "nature is a stock house" mindset, would we have laws and known ethics to account for these problems? I think so, but it is going to take a lot to change enough people's mindsets to make better rules.

Writing their response papers also offers a good opportunity for students to begin articulating their own opinions about how other beings should be treated. Students will often make comments like the following:

Student Writing: Response Papers

by Jennie Williams

People as well as plants, animals, and environments should be given equal rights to life and respect as humans selfishly believe that they have.

When students come to class, we discuss the readings and their reactions. The readings in this lesson are challenging, as they force students to face cruel realities of humankind's treatment of other species, to wrestle with extremely difficult ethical questions, and to reflect on the ethical implications of their own behavior. This can make students feel overwhelmed or defensive. It's worth talking through this. Students often react by saying that treatment of other species, whether ethical or not, will never change. While it's understandable to ask "how could any of this change?," raising this doubt shouldn't end the conversation.

To help get students thinking about how ethical questions are addressed, or if they are, in our society, I have students answer a set of questions, working in groups. First I give them the following general questions:

Ethics Questions, Part 1:

1. What does it mean to be ethical/moral?
2. What are the ethical values that we, as a society, live by?

3. What do we base our ethical decisions on in our daily lives? What do you, personally, base your ethical decisions on? What is the ultimate goal of our decisions?

4. How do we decide who deserves rights and who doesn't, and what rights they deserve? What do we base these decisions on?

I move through each group while they talk, and then as a full class we talk through their answers to these questions. I encourage students to consider what mainstream ethical norms exist in our society, how those came to be, and who benefits from those principles. It's important to include this discussion of *who benefits* from a society's ethical norms, and how a society decides what groups deserve rights and what groups don't. Power plays a part in the granting of rights, and that shouldn't be overlooked. I have students talk through examples of groups that have gained greater ethical consideration in the past – including women and African Americans, for example – and how they did so. I also ask students to consider how people's avowed ethical principles compare to their everyday actions – does people's behavior typically align with the principles they say they believe in?

Once we've discussed as a class, I have students return to their groups and answer a second set of questions:

Ethics Questions, Part 2:

Answer the following questions for each of these three authors: Peter Singer, Joan Dunayer, and Christopher Stone.

1. Summarize the author's argument. What is this author saying about who should have rights and what ethical values we should base our decisions on? Do you agree with her/his argument?

2. Give at least one specific example of how the author supports her/his argument, and discuss how effective you think the author's points are.

3. Select at least one quote that you think exemplifies this author's argument or illustrates a key point the author makes.

4. Imagine that we, as a society, committed to behaving as this author suggests we should. What sorts of changes would our society have to make? Give specific examples.

This set of questions should help students compare the details of the arguments made by each of these three authors, who all argue for different sorts of rights and ethical considerations for nonhuman beings. Make sure students draw on quotes and direct examples from the text to illustrate their conclusions about each author's proposals. For example, in this excerpt a student identifies one of the central points of Peter Singer's argument:

Student Writing: Response Papers

by Chelsie Bateman

He makes some valid points, one of which is, "The question is not, Can they reason? nor Can they talk? but, Can they suffer?" (Singer, Pg 6). This is like the basis of nonhuman equality. Of course they can suffer, just like us.

Singer argues that any being who suffers must have that suffering taken into account; this may not mean it's the only factor, but that it should be included in our reasoning. Dunayer argues that nonhuman animals should be free from being used and exploited for human purposes, should have their own legal rights, and should not be considered property. Stone argues for legal arrangements in which a human may serve as a guardian or trustee and act in court on behalf of ecosystems and natural features – like forests or streams – to defend their integrity and well-being.

Once again, as students answer this set of questions I spend time with each group to help guide them through their analysis of the authors' arguments. For the last question, I have them talk through what society would be like if it were aligned with the ethical principles described by each author. For example, students often suggest that if we operated as Dunayer suggests, all humans would be vegetarian. They also ask if having nonhuman animals as pets is permissible within Dunayer's ethical formulation. They generally conclude that, by Singer's approach, much animal testing would cease, but they suggest that the changes wouldn't be as sweeping as with Dunayer. Once students have talked through these questions in their groups, we return to whole-class discussion and they share what they discussed.

In some semesters I have students write down and submit their answers to these questions; in other semesters I simply tell them to make notes and sum up their conversations when they report back to the rest of the class. Students may not be sure if they agree with the authors' arguments, and they may not reach any firm conclusions about what is right and wrong, but this lesson should get them thinking about questions that are too often ignored.

Products and Assessment

The products that come from this lesson include students' written response papers, their participation in class discussion, and, if you choose to require it, their written answers to the assigned questions.

I use the following grading criteria to evaluate student work for this lesson. I have, in different classes, formatted these criteria as a grading rubric or listed them in my course syllabus. How to format and apply the grading criteria will depend on individual context, school requirements, and teacher preference.

Grading Criteria

- Student demonstrates critical thinking about ethical principles.
- Student analyzes the implications of diverse ethical principles using evidence from assigned texts.
- Student formulates informed opinions about the ethics of modern treatment of nonhuman beings, including practices such as animal testing, factory farming, and clear-cutting.
- Students reflects on personal behavior in light of ethical principles.
- Student demonstrates critical analysis of the arguments made by assigned authors, how they play out in their lived experiences, and their relevance to social and environmental justice.
- Work demonstrates thorough reading and comprehension of assigned course texts.
- Work demonstrates personal reflection, critical thought, and insight into course texts.
- Arguments are clear, well-developed, and documented with evidence from texts.
- Work demonstrates critical analysis of course topics, questions, and subject-matter.
- Style, usage, format, grammar, imagery, and presentation support meaning and are intentional, creative, and original.
- Work meets all requirements of the assignment and is utilized to facilitate development of personal understanding.

LESSON 7.2: WHO HAS RIGHTS

Legal Rights for Humans and Nonhumans in the U.S. and Abroad

About this Lesson

In this lesson students read legal documents from the U.S. and Ecuador to compare what beings are granted rights in different countries; they also read about legal battles to expand rights to nonhuman beings and arguments as to what rights should be granted to nonhuman beings and the land.

Lesson Objectives

Students will:

- Learn about legal rights afforded to humans and nonhumans in the U.S. and in other countries.
- Reflect on their own beliefs about who should be afforded rights and what rights should be granted.
- Compose original analytical writing.

Lesson Activities at a Glance

1. Students read assigned texts exploring legal rights for humans and nonhumans.
2. Students write response papers outside of class.
3. In-class discussion.
4. Students work in groups to develop lists of the rights they feel should be extended to nonhuman beings.

Key Content Area Skills: reading and analyzing nonfiction prose, analyzing legal documents, writing analytical and reflective essays.

Texts Used in this Lesson

Constitution of the Republic of Ecuador: Preamble and excerpt from the chapter "Rights for Nature" (see the online resource for text)

U.S. Constitution and U.S. Declaration of Independence, excerpts (see the online resource)

Vandana Shiva, *Earth Democracy*, 9–11

Charles Siebert, "Should a Chimp Be Able to Sue Its Owner?" *New York Times*

Note: Publication information for these readings is listed at the end of the unit. Sample passages are included in the full List of Readings at the end of the book.

My Procedure

To start, students read the assigned texts and write response papers outside of class. These readings explore legal rights for nonhuman beings, an issue under debate in many countries. The assigned texts include an excerpt from the Constitution of the Republic of Ecuador that establishes "Rights for Nature," an example of certain rights for nonhuman beings that have been established in a handful of other countries. Students are often struck by this excerpt, and express positive opinions about codifying rights for other beings. Here's an example:

Student Writing: Response Papers

by Michelle Ott

It is to my belief that society makes empty promises and takes minimal efforts to conserve the natural world. We educate our children about conservation; however, little action is taken. We do not differentiate between right and wrong. We do not call for change. We talk about what we can do, but we do not act. I feel as though we could make great strides if we were to model the efforts that the Republic of Ecuador has taken to create a, "new form of national coexistence, in diversity and harmony with nature, in order to achieve good living" (1). The Republic of Ecuador is a great example of how to take action. They have established a Constitution to include rights for nature and how they plan to preserve the natural world. We can no longer simply talk about conserving; we need to take action.

Students frequently write about the possible benefits of establishing more legal rights for nonhuman beings, as in this example where a student suggests that such formalized rights might encourage people to feel differently toward other beings:

Student Writing: Response Papers

by Jacob Rosenborough

I feel as if America would be much more respectful towards nature if it was given constitutional rights, it would put our relationship with nature into a different perspective that we would more readily understand and respect. . . . We do have laws that prevent animal abuse, pollution, and in some ways the destruction of nature, but these are not comprehensive enough.

Other students may also point out that the rights we grant – or don't grant – other species hold a mirror up to reveal our beliefs:

Student Writing: Response Papers

by Rebecca Postowski

Comparing the U.S. Constitution with Ecuador's was eye opening to me, as there were major value differences. . . . The reason why I think this comparison is so important is because these documents represent a country's values and beliefs. . . . The policies that we enact and produce for the environment reflect our attitudes towards the environment.

Another of the assigned readings is a passage by Vandana Shiva in which she lays out ten "principles of earth democracy." These principles establish rights for all living beings, as Shiva argues that all beings have intrinsic worth, deserve sustenance and compassion, and should be considered co-participants in community. This reading also sparks much discussion among students. Many students are inspired by Shiva's list, though just as many comment that they believe it would be unrealistic to enact. It's worth interrogating this question of whether Shiva's recommendations could be "realistic," and asking students what might stand in the way. Here's a great example of a student exploring that exact question:

Student Writing: Response Papers

by Jacob Rosenborough

I agree with her over all [*sic*] perception of what an "Earth Democracy" should look like, it seems fair, practical, and comprehensive on all levels of society (as she mentions "vibrant local economies" all the way up to the whole Earth). The only problem is that a lot of people in western society see this mindset as silly, saying that animals aren't as smart as us and that we own them. This clashing between these ideas only slows down progress to achieving rights for nonhuman animals.

When students come to class after reading these materials, I start by discussing their reactions. I ask what they think of the Constitution of Ecuador and of Shiva's recommendations, and what they think of the effort to establish more legal rights for nonhuman animals in the U.S., as discussed in the article by Charles Siebert.

Then I have students work with partners or in groups to develop their own list of the legal rights they believe should be granted to nonhuman animals and to ecosystems. Their reasoning should be drawn from the readings in Lesson 7.1 as well as those in this lesson. Students may adopt recommendations from any of the authors they've read in this unit and add or mix together different proposed rights. For example, students may conclude that all animals should have the right to live as they choose and not be caged, moved, or used for others' purposes. They may conclude that ecosystems have the right to maintain themselves, and therefore must have enough space, biodiversity, and health of the water and air. They may suggest that all animals should be granted legal personhood and should be allowed a legal proxy to represent them in court if they are harmed.

Once students have formed their lists, I have them share with the whole class and discuss. If you wish, you could also have students submit a written copy of their lists of rights. I don't expect students to use correct legal lexicons, but this exercise should get them thinking closely about their own beliefs and the implications of granting or not granting legal rights to living beings.

Products and Assessment

The products that come from this lesson are students' written response papers and their participation in class discussion. If you choose, you may also have students write and submit the list of legal rights they develop.

I use the following grading criteria to evaluate student work for this lesson. I have, in different classes, formatted these criteria as a grading rubric or listed them in my course syllabus. How to format and apply the grading criteria will depend on individual context, school requirements, and teacher preference.

Grading Criteria

- Student demonstrates critical thinking about ethical principles and legal rights.
- Student demonstrates ethical reflection and formulates informed opinions about what rights should be granted to nonhuman beings.
- Student demonstrates critical analysis of the arguments made by assigned authors, how they play out in their lived experiences, and their relevance to social and environmental justice.
- Work demonstrates thorough reading and comprehension of assigned course texts.
- Work demonstrates personal reflection, critical thought, and insight into course texts.
- Arguments are clear, well-developed, and documented with evidence from texts.
- Work demonstrates critical analysis of course topics, questions, and subject-matter.
- Style, usage, format, grammar, imagery, and presentation support meaning and are intentional, creative, and original.
- Work meets all requirements of the assignment and is utilized to facilitate development of personal understanding.

LESSON 7.3: ENVIRONMENTAL JUSTICE

Who Feels the Impact of Environmental Problems the Most and Why

About this Lesson

In this lesson students learn about the disproportionately negative effects of environmental degradation often experienced by marginalized groups, including people of color, low-income families, and people in less wealthy and less industrialized countries.

Lesson Objectives

Students will:

- Learn about the concept of environmental justice and about which people most frequently experience the impacts of pollution and environmental destruction.
- Critically analyze the social forces that result in environmental racism and other issues of environmental justice.
- Reflect on potential interventions to address issues of environmental justice.
- Compose original analytical writing.

Lesson Activities at a Glance

1. Students read assigned texts on environmental racism and environmental justice.
2. Students write response papers outside of class.
3. In-class discussion and video viewing.
4. If applicable, students learn about a local issue of environmental justice.

Key Content Area Skills: reading and analyzing nonfiction prose, critical analysis of social issues and dynamics, writing analytical and reflective essays.

Texts Used in this Lesson

Patrick Hossay, *Unsustainable: A Primer for Global Environmental Justice*, 1–41

Robert Bullard, "Anatomy of Environmental Racism," in *Environmental Discourse and Practice: A Reader*, 223–231

Will Heford, "That God Made," in *From Totems to Hip-Hop: A Multicultural Anthology of Poetry Across the Americas, 1900–2002*, 216–217

Juan Felipe Herrera, "Earth Chorus," in *From Totems to Hip-Hop: A Multicultural Anthology of Poetry Across the Americas, 1900–2002*, 30–32

Video

Chester Environmental Justice (see link in the online resource)

Note: Publication information for these materials is listed at the end of the unit. Sample passages are included in the full List of Readings at the end of the book.

My Procedure

Students start by reading the assigned texts for this lesson and writing response papers outside of class. Many of my students are unfamiliar with the concept of environmental racism or environmental justice at the start of this lesson. As one of my students writes:

Student Writing: Response Papers

by Christina Malliakos

I personally didn't know that race had anything to do with the environment. After I read this passage I realized that it does.

Students coming from more privileged environments might be uncomfortable or resistant when they first learn of environmental racism. When they learn that incinerators, landfills, and polluting factories are disproportionately built in neighborhoods occupied predominantly by lower-income people of color, my students are often hesitant to believe this still happens, or suggest that this situation must be more about class than race. In reality both race and class play major roles; studies indicate that in many circumstances race influences the issue regardless of class, while in other cases class is a significant factor as well (for studies, see Clark, Millet, and Marshall 2014; Smith 2014; Heiman 1996; Downey 1998). While the issue is complicated, students shouldn't shy away from the racial component. Help students work through any defensiveness or hesitation and focus on critically analyzing why this situation happens and what can be done about it.

When students come to class, I show the video *Chester Environmental Justice*, a 9-minute video containing footage from a 1996 documentary titled *Laid to Waste*. The video follows issues of environmental racism in Chester, Pennsylvania, including a fight over a proposed incinerator. I discuss both the video and the readings with students, working through their reactions and questions. I also remind students that environmental justice is an issue for people all over the U.S. and in other countries as well, since corporations and governments extract resources from poorer and "less developed" countries, build massive factories in these countries, and then often send toxic materials and waste to these countries for disposal.

After this discussion, I introduce students to a local issue of environmental justice. Depending on your location, you may be able to easily find a local issue in which a community is fighting to prevent some form of pollution from entering their neighborhood. This could be a proposed incinerator or landfill, a factory, natural gas drilling or "fracking," an oil pipeline, or a factory farm (these produce enormous amounts of waste runoff), for example. In my state of Maryland, an incinerator has been proposed in a lower-income neighborhood and a group of high school students has mounted a strong campaign to halt the project. I have had my college students interview these high school students, in one case creating a short documentary about the group. Depending on the issue and what stage the fight against it is in, you may have students research the issue, write reports on the neighborhood in question, interview residents, speak with law-makers about why this neighborhood has been chosen as the site for whatever the project may be, or research the possible negative impacts of the project. Because this project is so dependent on the specifics of your region and issue, I leave my recommendations open-ended. The key is to help students connect the subject of environmental racism and environmental justice to real people living in a real place in their own city or state, and to learn about what those people will or do experience as a result of living near polluting facilities or businesses.

Products and Assessment

The products that come from this lesson are students' written response papers and their participation in class discussion. Your students may also produce a product from their exploration of a local issue of environmental justice – this may include an interview, class presentation, research report, or short documentary, for example.

I use the following grading criteria to evaluate student work for this lesson. I have, in different classes, formatted these criteria as a grading rubric or listed them in my course syllabus. How to format and apply the grading criteria will depend on individual context, school requirements, and teacher preference.

Grading Criteria

- Student demonstrates understanding of the concept and process of environmental racism, and the social forces that produce conditions of environmental injustice.
- Student demonstrates critical thinking about how to address social and environmental justice issues for marginalized communities in the U.S. and globally.
- Student demonstrates knowledge of a local issue of environmental justice and its implications for people in the affected neighborhood or community.
- Student demonstrates critical analysis of the arguments made by assigned authors, how they play out in their lived experiences, and their relevance to social and environmental justice.

- Work demonstrates thorough reading and comprehension of assigned course texts.
- Work demonstrates personal reflection, critical thought, and insight into course texts.
- Arguments are clear, well-developed, and documented with evidence from texts.
- Work demonstrates critical analysis of course topics, questions, and subject-matter.
- Style, usage, format, grammar, imagery, and presentation support meaning and are intentional, creative, and original.
- Work meets all requirements of the assignment and is utilized to facilitate development of personal understanding.

EFFECT OF THE LESSONS: INFORMED ETHICAL THINKING

I believe it is vital for our society to actively engage in ongoing ethical dialogue, challenging ourselves to consider the motivations and ramifications of our behavior from ethical perspectives. This ethical dialogue must include consideration for the people, other animals, and ecosystems who are suffering as a result of our individual and collective actions. This sort of discussion already takes place among academics in fields like environmental ethics, ecofeminism, and feminist ethics (see, for example, Merchant 2000; Haraway 2003; Midgley 1984; Noddings 1984; Peterson 2009; Smith 2001; Stone 2010). But while these scholarly examinations continue, the insights they raise seem to barely register in mainstream discourse. Mick Smith comments that, "Some . . . do not even recognize the possibility of an environmental ethics. . . . They are content to continue to view the nonhuman world as of only instrumental value and to evaluate it using cost–benefit analyses or other economistic tools" (Smith 2001, 15).

By exposing students to ethical analyses, we can equip them to engage in their own ethical reflection and dialogue, and to recognize that such dialogue must involve those most harmed and least represented in dominant discourse. The readings in this unit each provide insights and innovative possibilities for how we can and should treat other beings. Considering these sorts of arguments, and developing strategies that could be used to employ them, should not be an optional or peripheral feature of society, but central to our cultural and educational efforts.

By the end of this unit students may not have reached any definitive conclusions about what they believe is right or what should be done to address the injustices explored in these lessons, but they should be honestly grappling with the ideas, issues, and proposals they're presented with. Throughout the unit students should be asking tough questions about what is right and wrong and what behaviors we can or should justify. Here's an example of a student asking complex and important questions about the rights of nonhuman animals and the land:

Student Writing: Response Papers

by Heather Harshbarger

Where is the line that it's appropriate to divide what is "ours" and what is "theirs," when they have no true way of their own to claim it? That's just something to think about when we think about our lives and how they connect to, or don't connect to, nature and other organisms/things.

I hope students will also notice, as Heather does in this excerpt, that *power* is involved in the claiming and granting of rights. Those who have little or no power to demand rights for themselves are often exploited. Here's another example of a student reflecting on the powerless position we have put other species in, a position in which decisions are made without their participation:

Student Writing: Response Papers

by Malarie Novotny

I think about animals the most, however, just because I see a ton of roadkill. It makes me sad that they are just trying to live naturally, without means of an "animal crossing" kind of system, a language they understand, and we really don't go out of our way to cohabit.

Throughout this unit students should be asking themselves who "should" have rights, what the criteria for determining rights should be, and how the well-being of other people and other creatures should be safeguarded. Here's a student asking important questions about what rights plants should have:

Student Writing: Response Papers

by Chelsie Bateman

Now, what about plants? Do they have rights as well? Do they suffer? I cannot say for a fact that they do, but they should have rights that involve preventing us from exploiting and degrading them and their habitats.

Students should be also questioning what we *owe* to other beings, what responsibilities we have for others and how we should live together in ways that are mutually positive:

Student Writing: Response Papers

by Heather Harshbarger

Leopold, in A Sand County Almanac, makes a good point that seems to flow with the rest of the readings. He states "In human history, we have learned that the conqueror role is eventually self-defeating" (240). In a way, it may be saying that humans are destroying themselves by trying to redirect and take over the land. However, it also says we feel no true obligation to what surrounds us through [*sic*] "The land-relation is still strictly economic, entailing privileges but not obligations" (238). We own the land, which we readily acknowledge. What we seldom show is that we may have something we owe to the land, or that we should be taking care of it. By trying to own it without having any true relation to it, we are "conquering" it in a way.

They may also start to highlight relationships and mutual participation as key factors for forging ethical, just modes of coexistence. Here a student responds to the idea that our sense of "community," and our participation as ethical members of a community, should include the more-than-human world:

Student Writing: Response Papers

by Jennie Williams

People have a responsibility to involve nature including our environment, plants, and animals in our commitments to our community.

Valuing relationships and personal feelings of care should be a key part of these discussions of ethics. Ethics isn't just cold theories, it's also very personal – our desire to protect the well-being of others comes, in part, from our desire to protect those we love. In this way ethics is simply about expanding the circle of those we love to include the larger world. Ethicist Anna Peterson (2009) points out the very personal origins of ethical values, suggesting that ethical behavior isn't as "difficult and rare" as we may sometimes assume (133). She says:

> Few people experience care for children or protection of beloved places as sacrifices or even as deliberate choices. Rather, such actions feel natural, even necessary, since they emerge from core aspects of our personal histories and identities. Everyday experiences and relationships engender commitments and loyalties that translate into ethical values: commitments to something larger than our own self-interest.
>
> *(Peterson 2009, 134)*

Peterson suggests that we need to find ways to include these personal commitments to people and places we love in our thinking about ethical questions in the public sphere. What we love personally and what we choose to protect as a society should not be separate questions, and attending to ethical questions should mean exploring and cultivating love for others.

Exploring ethical questions also means finding ways to give a voice to those who have been excluded from dialogue. This includes marginalized peoples, nonhuman animals, plants, and ecosystems. Seriously considering proposals for how to do this should also be a key facet of this unit. Whether it's establishing new laws, allowing humans to legally represent nonhumans in court, our applying principles of "earth democracy" as discussed by Vandana Shiva (in her text assigned in Lesson 7.2), it's important for students to think about how unethical systems can be changed. Here a student comments on Shiva:

Student Writing: Response Papers

by Chelsie Bateman

And speaking of rights, every creature on Earth has natural rights that we can't take away, and should not even try to take away. "No state or corporation has the right to erode or undermine these natural rights . . . " (Shiva, Pg 3–4). Like Vandana Shiva says, "We all have a duty to live in a manner that protects the earth's ecological processes, and the rights and welfare of all species and all people" (Shiva, Pg 3).

To protect the "rights and welfare of all species and all people" we will need to transform some aspects of our society. So, to think about ethical questions means thinking about how to change society. The sorts of conversations I hope students have during this unit can provide "the possibility of ethical critiques of current social relations and ethical arguments for social change" (Smith 2001, 17). Ethical dialogue points out essential questions we must ask ourselves about our behavior and choices, and also generates mechanisms we can use to critically re-envision our social structures and reformulate sustainable approaches to the world.

UNIT 7 READINGS AND REFERENCES

Bowen, Kevin. 2001. "Gelatin Factory." In *Poetry Like Bread*, edited by Martín Espada, 68–69. Willimantic, CT: Curbstone Press.

Bullard, Robert. 2000. "Anatomy of Environmental Racism." In *Environmental Discourse and Practice: A Reader*, edited by Lisa M. Benton and John Rennie Short, 223–231. Malden, MA: Wiley-Blackwell.

Chester Environmental Justice. 2008. Video recording. www.youtube.com/watch?v= 5Opr-uzet7Q&feature=youtube_gdata_player.

Clark, Lara P., Dylan B. Millet, and Julian D. Marshall. 2014. "National Patterns in Environmental Injustice and Inequality: Outdoor NO2 Air Pollution in the United States." *PLoS ONE* 9 (4): e94431. doi:10.1371/journal.pone.0094431.

"The Declaration of Independence." 2010. *Ushistory.org, Independence Hall Association.* www.ushistory.org/declaration/document/index.htm.

Downey, Liam. 1998. "Environmental Injustice: Is Race or Income a Better Predictor?" *Social Science Quarterly* 79 (4): 766–778.

Dunayer, Joan. 2001. *Animal Equality: Language and Liberation.* Derwood, MD: Ryce.

"Ecuador Constitution, 2008." 2009. *Political Database of the Americas.* http://pdba. georgetown.edu/Constitutions/Ecuador/ecuador08.html.

Haraway, Donna Jeanne. 2003. *The Companion Species Manifesto: Dogs, People, and Significant Otherness.* Chicago: Prickly Paradigm Press.

Heford, Will. 2003. "That God Made." In *From Totems to Hip-Hop: A Multicultural Anthology of Poetry Across the Americas, 1900–2002*, edited by Ishmael Reed, 216–217. New York: Thunder's Mouth Press.

Heiman, Michael. 1996. "Race, Waste, and Class: New Perspectives on Environmental Justice." *Environmental Justice Net*. April. www.ejnet.org/ej/rwc.html.

Herrera, Juan Felipe. 2003. "Earth Chorus." In *From Totems to Hip-Hop: A Multicultural Anthology of Poetry Across the Americas, 1900–2002*, edited by Ishmael Reed, 30–32. New York: Thunder's Mouth Press.

Hossay, Patrick. 2006. *Unsustainable: A Primer for Global Environmental and Social Justice*. London: Zed Books.

Leopold, Aldo. 1970. *A Sand County Almanac: With Essays on Conservation from Round River*. 1st Ballantine Books ed. New York: Ballantine Books.

Lorca, Federico García. 1980. "New York (Office and Attack)." In *News of the Universe: Poems of Twofold Consciousness*, edited and translated by Robert Bly, 110–112. San Francisco: Sierra Club Books.

Merchant, Carolyn. 2000. "Ecofeminism." In *Environmental Discourse and Practice: A Reader*, edited by Lisa M. Benton and John Rennie Short, 209–213. Malden, MA: Wiley-Blackwell.

Midgley, Mary. 1984. *Animals and Why They Matter*. Athens: University of Georgia Press.

Noddings, Nel. 1984. *Caring: A Feminine Approach to Ethics and Moral Education*. Berkeley: University of California Press.

Peterson, Anna Lisa. 2009. *Everyday Ethics and Social Change: The Education of Desire*. New York: Columbia University Press.

Shiva, Vandana. 2005. *Earth Democracy: Justice, Sustainability, and Peace*. Cambridge, MA: South End Press.

Siebert, Charles. 2014. "Should a Chimp Be Able to Sue Its Owner?" *New York Times*. April 23. www.nytimes.com/2014/04/27/magazine/the-rights-of-man-and-beast.html.

Singer, Peter. 2002. *Animal Liberation*. New York: Ecco.

Smith, Mick. 2001. *An Ethics of Place: Radical Ecology, Postmodernity, and Social Theory*. Albany, NY: State University of New York Press.

Smith, S. E. 2014. "Environmental Racism Is Consistent Across Economic Classes." *This Ain't Livin'*. July 14. http://meloukhia.net/2014/07/environmental_racism_is_consistent_across_economic_classes/.

Stone, Christopher D. 2010. *Should Trees Have Standing? Law, Morality, and the Environment*. 3rd ed. New York: Oxford University Press.

"U.S. Constitution." 2010. *Legal Information Institute*. Accessed August 31. http://topics.law.cornell.edu/constitution.

Unit 8

IMAGINING POSSIBLE FUTURES

Consider the cherry tree: thousands of blossoms create fruit for birds, humans, and other animals, in order that one pit might eventually fall onto the ground, take root, and grow. Who would look at the ground littered with cherry blossoms and complain, "How inefficient and wasteful!" The tree makes copious blossoms and fruit without depleting its environment. Once they fall on the ground, their materials decompose and break down into nutrients that nourish microorganisms, insects, plants, animals, and soil. Although the tree actually makes more of its "product" than it needs for its own success in an ecosystem, this abundance has evolved . . . to serve rich and varied purposes. In fact, the tree's fecundity nourishes just about everything around it.

What might the human-built world look like if a cherry tree had produced it?

We know what an eco-efficient building looks like. It is a big energy saver. It minimizes air infiltration by sealing places that might leak. (The windows do not open.) It lowers solar income with dark-tinted glass, diminishing the cooling load on the building's air-conditioning system . . .

Here's how we imagine the cherry tree would do it: during the daytime, light pours in. Views of the outdoors through large, untinted windows are plentiful. . . . The windows open. The cooling system maximizes natural airflows, as in a hacienda: at night, the system flushes the building with cool evening air, bringing the temperature down and clearing the room of stale air and toxins. A layer of native grasses covers the building's roof, making it more attractive to songbirds and absorbing water runoff . . .

> If nature adhered to the human model of efficiency, there would be fewer cherry blossoms, and fewer nutrients. . . . Fewer songbirds. Less diversity, less creativity and delight. The marvelous thing about effective systems is that one wants more of them, not less.
>
> —*McDonough and Braungart (72–77)*
> *Lesson 8.2*

The lessons in this unit ask students to respond to disparate visions of the future and to explore proposed changes to our modern ways of living, including new forms of engineering and building and new systems for housing, transportation, and food production. Assignments engage students in creative imagining of new future paths and require students to analyze what changes in values and policies would be necessary to achieve such alternate paths.

Some key questions that the lessons in this unit should raise for students:

- What do you want the world to be like in the future? How do we make that future world come about?
- How might we do things differently than we do today in order to achieve a healthier, more just, more sustainable future?
- What are some possibilities for living, building, transporting, and feeding ourselves differently? What might the world be like if we tried these alternatives?

LESSON 8.1: IMAGINING THE FUTURE

Views of Possible Futures in Literature and Film

About this Lesson

In this lesson students read and view narratives of possible futures in literature and film. They consider these imagined views of what the future could be like and they draw conclusions, based on current human behaviors, as to what aspects of these imagined futures could come to pass.

Lesson Objectives

Students will:

- Reflect on possible future paths as described in literature and film.
- Conduct research on current events.
- Reason about possible futures based on knowledge of current events.
- Critically analyze literature and film.

Lesson Activities at a Glance

1. Students read assigned texts describing possible futures.
2. Students write response papers outside of class.
3. In-class discussion.
4. Students work in groups to view a film outside of class and create a presentation about the film.

Key Content Area Skills: reading and analyzing nonfiction prose, viewing and analyzing film, writing analytical and reflective essays, creating multimedia presentations.

Texts Used in this Lesson

Passages from *Reading the Environment*, edited by Melissa Walker:

"A Fable for Tomorrow" by Rachel Carson, 523–524
"Two Children in a Future World" by George Mitchell, 525–528

Ernest Callenbach, *Ecotopia*, 1–28

Films

(Students select one to view outside of class)

> *The Day After Tomorrow*
> *WALL-E*
> *Mad Max 2*
> *Idiocracy*
> *Soylent Green*

Optional: *Snowpiercer* (this film may be too violent for some students, use your judgment in choosing whether to assign it)

Note: Publication information for these materials is listed at the end of the unit. Sample passages are included in the full List of Readings at the end of the book.

My Procedure

Students start by reading the assigned texts for this lesson and writing response papers outside of class. These readings are fictional descriptions of possible futures— the stories by Carson and Mitchell are dystopic, while the selection from Ernest Callenbach's *Ecotopia* describes a more idyllic future. *Ecotopia* is a novel narrated by a visitor to an environmentally sustainable society called "Ecotopia," and as an outsider this narrator starts off skeptical of the lifestyle he witnesses. In this way he serves to be more aligned with a typical reader. Over the course of the text his perspective changes, and he eventually comes to like this society and wants to stay. You may need to help students understand the narrative strategies at work in this text. Here's an example of a student commenting on it:

Student Writing: Response Papers

by Chelsie Bateman

At first I was like, Where is this Ecotopia? because I actually thought it was real, but then as I continued to read, it became a cool story and a place that I wanted to visit. . . . I think that using some of the techniques that Ecotopia uses in today's society would give a foundation to the start of us becoming more sustainable.

As well as reacting to the ideas in *Ecotopia*, students are also quick to comment on the negative possibilities described by Carson and Mitchell. Here's an anonymous student discussing Mitchell's story, and raising the idea that "the future" isn't some distant point, but is something we should work to influence every day:

Student Writing: Response Papers

If life is "expected to be an endless supply of problems" (529) we shouldn't hesitate any longer to try to find solutions. Time is continuous and we can never fully say when we will reach the future, so the [*sic*] each day should be invested in making the present more environmentally healthy and sustainable because the present eventually builds into a future.

After students have read these texts and written their response papers, I give them an assignment in which they view films that are set in the future and create presentations about the films. For this assignment I have students work in groups. Each group selects one of the films listed above (or students sign up for one of the films, and this arranges them into groups). They watch the film outside of class, and do research on current environmental conditions related to the possible futures presented in the films. Then they give a presentation in class. Here is the assignment:

Assignment: Film Report on Possible Futures

In class, you'll be given a list of films. Sign up to view one of these films and research the issues raised in the film. Then, in small groups, prepare a presentation about the film.

Your presentation should include the following components:

1. Describe the film. Identify the genre and year the film was made. Then outline the plot. You may choose to show a short clip or images from the film during your summary.
2. What is the message of the film and what did you learn? Do the filmmakers have a specific opinion and point of view they are trying to convey? If so, what is it? How can you tell? Does the film have a 'moral'? What information did you learn by watching it?
3. Is this the future we're heading for? Consider the events portrayed in the film. How did they come about? What environmental conditions led to the state of the world in this future scenario? Now consider these conditions as they are occurring today. Do some research, and report on what is happening now. Make sure you use reliable sources and cite all references. Then tell us whether you think that current conditions could eventually lead to any of the events in the film.
4. Your response to the film. Tell us what you thought of the film. Did you feel aligned with the perspective of the filmmaker? Did you identify with any of the characters? Why/why not? How did you feel after the film ended? Did it change your attitudes in any way? Are you worried about these events occurring in the future?

Your presentation should be 10–15 minutes in length. Submit notes and full citations in your presentation file or as a separate document accompanying your presentation.

If, for example, students watch *WALL-E*, then in their research for this presentation they might investigate how many pounds of garbage enter landfills every year, how many pounds the average American throws away in a week or a month, where landfills are located and how big they are, how much garbage goes into oceans each year, how much plastic can be found in the Pacific trash vortex, and the like. They should include this research in their presentation, and use the information to draw conclusions about whether they think humans will ever fill the planet with trash as seen in *WALL-E*. (For this film students could also research corporate influence in politics, the number of advertisements an average American views per day, obesity rates, and other such issues also explored in the film.) Students watching *The Day After Tomorrow* would research global climate change, levels of CO_2 currently being emitted into the atmosphere, and the most current predications from climate scientists as to how much sea level is likely to rise and what other changes are likely to occur as humans continue to alter the atmosphere. For *Mad Max 2*, students may research current levels of petroleum usage, dependence on cars, effects of oil drilling, peak oil, and predictions as to how long fossil fuels (as a non-renewable resource) will last. For *Soylent Green*, students should not focus on whether people might someday eat other humans, but should focus on researching topics like food security, corporate control of food, and conditions that affect the supply of fruits and vegetables – this should include research on bees and colony collapse disorder, changing weather patterns and levels of rainfall due to climate change, and soil depletion.

After students give their presentations in class, we discuss what everyone took away from the films and the presentations. I ask students which possible future scenarios they find most worrisome and why. We then create a list together of the major problems they feel must be addressed in order to avoid such negative future paths. I ask students to list all significant challenges to a sustainable and just future, and I type the list live as they make suggestions, projecting it onscreen from my computer. In the bow below you'll find the exact list we made in one of my semesters, typed directly from my students' suggestions.

Class List: Problems/Challenges to Sustainability and Justice

- oil – dependency, access, limited quantity remaining, expense
- overconsumption – buying, waste, overproduction – resource-intensive processes
- waste – proper disposal of materials, quantity of waste
- awareness – public knowledge of problems
- climate, greenhouse emissions, severe weather patterns
- habitat destruction
- deforestation – clearing for industrial agriculture, industrial meat, building

- overgrazing – agricultural practices, quantity of meat consumption in western world
- health effects from pollution – asthma, acid rain, cancers, toxins
- population growth
- cruel treatment of nonhuman animals, lack of respect – policy, abuse, cosmetic testing, food industry
- food industry – inhumane treatment of animals, regulations that support big corporations, access to healthy/organic foods, subsidies and price of different foods, high consumption of meat – resource intensive, agriculture based on pesticides, runoff, labor conditions for food workers
- water pollution, shortages, water contamination, water rights, reduction in topsoil, water waste
- GMOs – nutrition/health content questions, patents prevent farmers from saving seed, corporate control of seed, public knowledge
- cruel treatment of humans, lack of respect – child labor, trafficking, poor populations facing pollution/exploitation
- urban/suburban sprawl – car pollution, permeability, heat island, car-dependent, fuel use
- power of corporations – influence on policy, government corruption, revolving door
- waste management – location of waste, waste in oceans
- human growth/living space – land destruction leading to people moving to slums
- mass extinctions, harm to ecosystems from loss of species diversity

This is a lengthy list! (Note that I teach this lesson at the end of my semester, and students have by this time learned about a great many environmental and social justice issues, as reflected in the comprehensive list they were able to create.) My students sometimes express feeling somewhat depressed and overwhelmed after creating this list, looking at how many challenges we face. In Lesson 8.3 we use this list again, creating a companion list of solutions that will address these problems. If you don't plan to teach Lesson 8.3, I recommend following this list by making a list of possible solutions. Understanding what conditions we might face in the future if we don't change our behavior is very important, but it should be paired with creative thinking about how to change the future. This is what Lessons 8.2 and 8.3 aim to provide.

Products and Assessment

The products that come from this lesson are students' written response papers, their participation in class discussion, and their film presentations. Presentations should show thoughtful analysis of the film and also effective research into environmental conditions related to those described in the film.

I use the following grading criteria to evaluate student work for this lesson. I have, in different classes, formatted these criteria as a grading rubric or listed them in my course syllabus. How to format and apply the grading criteria will depend on individual context, school requirements, and teacher preference.

Grading Criteria

- Student demonstrates critical analysis of literary prose and film.
- Student conducts research into environmental conditions using reliable sources and cites all sources properly.
- Student demonstrates thoughtful reflection on the possible long-term outcomes of current human practices, especially relating to the condition of the natural world.
- Student demonstrates critical analysis of the arguments made by assigned authors, how they play out in their lived experiences, and their relevance to social and environmental justice.
- Work demonstrates thorough reading and comprehension of assigned course texts.
- Work demonstrates personal reflection, critical thought, and insight into course texts.
- Arguments are clear, well-developed, and documented with evidence from texts.
- Work demonstrates critical analysis of course topics, questions, and subject-matter.
- Style, usage, format, grammar, imagery, and presentation support meaning and are intentional, creative, and original.
- Work meets all requirements of the assignment and is utilized to facilitate development of personal understanding.

LESSON 8.2: MAKING NEW FUTURES

Strategies for Just and Sustainable Future Paths

About this Lesson

In this lesson students read and view proposed strategies to achieve more sustainable and just future paths. They discuss these ideas for more sustainable modes of living, and analyze their possible impact on the future.

Lesson Objectives

Students will:

- Learn about alternative modes of building, engineering, and food production that may be more sustainable than contemporary processes.
- Reflect critically on the advantages and implications of proposals for sustainable modes of living.
- Compose original analytical writing.

Lesson Activities at a Glance

1. Students read assigned texts that explore ideas for how to live differently.
2. Students write response papers outside of class.
3. Film viewing in class about Earthships, an alternative way of building homes.
4. In-class discussion.

Key Content Area Skills: reading and analyzing nonfiction prose, writing analytical and reflective essays.

Texts Used in this Lesson

William McDonough and Michael Braungart, *Cradle to Cradle: Remaking the Way We Make Things*, 68–83, 102–105, 114–115

Passage from *Reading the Environment*, edited by Melissa Walker:

"Change the Way We Think: Actions Will Follow" by Bill McKibben, 567–571

Vandana Shiva, *Earth Democracy*, 9–11

Toby Hemenway, *Gaia's Garden*, 21–31, 120–123, 208–212

Optional: William McDonough and Michael Braungart, *Cradle to Cradle,* 118–156

Optional: Marek Oziewicz, "'We Cooperate, or We Die': Sustainable Coexistence in Terry Pratchett's The Amazing Maurice and His Educated Rodents," *Children's Literature in Education* 2009

Film

Garbage Warrior

Note: Publication information for these materials is listed at the end of the unit. Sample passages are included in the full List of Readings at the end of the book.

My Procedure

Students read the assigned texts for this lesson and write response papers outside of class. These texts include innovative ideas for living in ways that don't harm people, other beings, or the planet. Students generally very much enjoy the readings in this lesson. Many comment on the recommendations made by McDonough and Braungart, who argue that instead of trying to design and manufacture goods that simply "pollute less" and "minimize harm" to people and the planet, we should be designing materials that actually benefit the planet and can always either be reused or become part of a biological life-cycle. In this way materials would never be "thrown away" but would contribute to future materials or decompose and nourish the soil. Here's a student discussing some of the ideas from this reading:

Student Writing: Response Papers

by Jacob Rosenborough

"Cradle to Cradle" presented an idea alternative to what the usual environmentalist mindset holds as how to be green. Distinguishing between "efficiency" and "effectiveness" shows a way of thinking about the situation completely differently, while still achieving a sustainable system that makes sense for everyone. This "cradle-to-cradle" system doesn't try to reduce waste; but recognize that waste is a part of nature. With this in mind, humans only make things that produce productive waste. As I've said this is somewhat detached from what we think of as being green. One of the more common or popular expressions of being environmentally friendly is "reduce your carbon footprint". "Reducing" implies the idea of less waste and efficiency. Trying to force this word and idea in solving our climate, energy, and resource problems does nothing but hold us back from a more practical system that we should be striving towards instead.

The recommendations in this text are a great way to get students thinking differently about how the world *could* work. It's very important to break out of thinking that the way we currently do things is the only way they could be done. Here another student comments on this reading:

Student Writing: Response Papers

by Jennie Williams

I really liked the image of a cherry tree and how although when looking at the "ground littered with cherry blossoms" it is not "inefficient and wasteful!" (73). The cherry tree makes ample amounts of cherries, and natural resources that are not damaging the surrounding environment. Everything the tree drops is decomposed and returned naturally to the earth. If only our society created resources that could leave no trace of it's existence once used and disposed.

Imagining what humankind could do differently and envisioning better ways of being in the world proves very thought-provoking for students. When they come to class after reading the assigned texts, I show the film *Garbage Warrior* in class. It profiles architect Michael Reynolds, the creator of "Earthships." Earthships are self-contained homes made from recycled and sustainable materials that supply their own power, water, heat, and to some extent food. Students are generally very struck by the film. We have come to hold an extremely standardized vision of how to build homes and supply them with things like water and power, and this film offers an evocative alternate vision.

After viewing the film, I discuss with students, asking them about their reactions to both the film and the readings. Then I ask them about their favorite ideas from the readings and the film. We discuss what specific innovations students find most interesting or exciting and why. Finally, I ask students how they think the world would be different if we put these ideas into place on a broad scale. What are the possible outcomes of these changes? What are the benefits? What would it take to make these changes happen? Discussing these alternative ideas can be inspiring and hopeful for students, and is an important precursor to Lesson 8.3, in which students create their own visions of possible futures.

Products and Assessment

Students' written response papers and their participation in class discussion are the products that can be used to evaluate achievement of the objectives for this lesson.

I use the following grading criteria to evaluate student work for this lesson. I have, in different classes, formatted these criteria as a grading rubric or listed them in my course syllabus. How to format and apply the grading criteria will depend on individual context, school requirements, and teacher preference.

Grading Criteria

- Student demonstrates critical analysis of proposals for alternative modes of living.
- Student thoughtfully reflects on the advantages and implications of proposed methods of building homes, manufacturing goods, and producing food, evaluating their potential impact on environmental sustainability.
- Student demonstrates critical analysis of the arguments made by assigned authors, how they play out in their lived experiences, and their relevance to social and environmental justice.
- Work demonstrates thorough reading and comprehension of assigned course texts.
- Work demonstrates personal reflection, critical thought, and insight into course texts.
- Arguments are clear, well-developed, and documented with evidence from texts.
- Work demonstrates critical analysis of course topics, questions, and subject-matter.
- Style, usage, format, grammar, imagery, and presentation support meaning and are intentional, creative, and original.
- Work meets all requirements of the assignment and is utilized to facilitate development of personal understanding.

LESSON 8.3: REIMAGINING THE FUTURE

Envisioning Our Own Ideas of a Better Future

About this Lesson

In this lesson students draw on the ideas they've discussed in previous lessons to creatively envision alternative future paths.

Lesson Objectives

Students will:

- Creatively envision the future to imagine alternative modes of living.
- Develop strategies for producing sustainable and just conditions that support the life and well-being of humans, nonhumans, and ecosystems.

Lesson Activities at a Glance

1. In-class discussion.
2. Students draw and write ideas for possible futures.
3. Students work together to build a list of possible solutions to problems of environmental and social justice.

Key Content Area Skills: writing creatively, imaginative problem-solving.

Texts Used in this Lesson

There are no new texts for this lesson, but it builds on the readings from Lessons 8.1 and 8.2.

My Procedure

This lesson follows on from the ideas and visions for the future explored in Lessons 8.1 and 8.2. To start this lesson, review those previous readings and discussions if necessary. I also review the list of challenges to sustainability students created in Lesson 8.1, putting it back up onscreen for students to read over.

Next I have students get into groups. I ask them to work together to imagine two possible futures: one negative and one positive. For the negative future, I tell them to imagine what they think the world will be like if people continue with the destructive and unjust practices we have discussed in previous lessons. What conditions will these behaviors eventually lead to? For the positive future, I ask students to think about what a sustainable and just world would look like.

I have students talk in detail about what they think each of these two futures might be like. In these futures, how do people live? What do their homes look like, how do they acquire food, what sort of transportation do they use? I encourage students to think about the suggestions made in previous readings as they answer these questions, and I also allow them to use laptops or other devices to do internet searches that might spark ideas for them – I suggest doing Google image searches using terms like "green building," "sustainable building," and "urban farming." The goal of these searches is just to encourage creativity and stimulate students' imaginations.

As students develop their visions of these two futures, I have them draw pictures and take notes to describe each future. I don't require complex drawings, but I ask students to use some combination of words and sketches to help convey their visions. This assignment could easily be developed to include a more involved artistic component, or you could have students write short stories describing the two futures they envision. I have students do their sketching and writing in class during one class period, but it could be turned into a longer assignment as well.

Once each group has talked through the details of their two visions for the future, and have made notes and drawings to help illustrate what they imagine, I have them share with the class. Each group describes and shows drawings of their negative vision of the future (these are usually images of a wasteland in which little or no life can survive), and then describes and shows drawings of their visions for a sustainable future. I ask questions about the details of their visions as students share.

Once each group has shared their visions of the future, I again bring up on the projector the list of challenges to sustainability and justice that we made as a class in Lesson 8.1. We review the list, and then I ask students to list suggestions for solutions to these problems. I type this new list on my computer (projected onscreen in the classroom) as they contribute to it. The box below contains the list my class created the same semester as the list quoted in Lesson 8.1, reprinted exactly as I typed it in class.

Class List: Possible Solutions for Achieving Sustainability and Justice

- subsidies only for maintaining environmental standards, not oil or corn
- subsidies for local farming
- urban gardening
- pollution tax
- govt support of research into green energy
- expand public transportation
- support for polyculture farming and plant-based cultures
- regulating against govt positions held by corporate employees
- stop revolving door
- campaign reform

- regulate against corporate funding of politicians
- stop patenting of life forms
- incentives and disincentives for waste disposal
- reduce food chains, increase local food providers
- govt regulations to reduce new construction
- values change away from buying unnecessary items
- financial disincentives away from purchasing products made unsustainably or at a distance
- less on shelves – feedback loops on how much resources are available and limits
- paperless schools
- price of goods based on full cost of production and disposal
- individual buying practices
- shift away from value of GDP
- localize for greater role in system
- sustainable urban design
- legal rights of nature

Once we've created this list, I ask students to think back over all of their ideas about how to create a healthy, sustainable, and just future, and then I ask them, one at a time, to share what they think the first steps must be in order to achieve those ideas. That is, what is the first essential change that must be made in order to transform our world into what we want it to be? We go around the classroom, and each student shares their answer to this question. Many say education is the first and most essential step. As we end this lesson, students should think about what they can do to take these first steps toward creating a world we all will truly want to live in.

Products and Assessment

Students' written response papers and their participation in class discussion are the products that can be used to evaluate achievement of the objectives for this lesson. You may also choose to collect their notes and drawings from their class activity envisioning possible futures.

I use the following grading criteria to evaluate student work for this lesson. I have, in different classes, formatted these criteria as a grading rubric or listed them in my course syllabus. How to format and apply the grading criteria will depend on individual context, school requirements, and teacher preference.

Grading Criteria

- Student demonstrates critical analysis of ideas for alternative modes of living.

- Student thoughtfully reflects on the advantages and implications of proposed methods of building homes, manufacturing goods, and producing food, evaluating their potential impact on environmental sustainability.
- Student shows creativity and innovative problem-solving in imagining alternative future paths.
- Work demonstrates thorough reading and comprehension of assigned course texts.
- Work demonstrates personal reflection, critical thought, and insight into course texts.
- Arguments are clear, well-developed, and documented with evidence from texts.
- Work demonstrates critical analysis of course topics, questions, and subject-matter.
- Style, usage, format, grammar, imagery, and presentation support meaning and are intentional, creative, and original.
- Work meets all requirements of the assignment and is utilized to facilitate development of personal understanding.

EFFECT OF THE LESSONS: IMAGINATION AND CREATIVE TRANSFORMATION

In order to create a better world we must be able to envision a better world. Therefore, one of the most essential skills that education can help to cultivate in students is the ability to imagine possibilities that are different from those presented by dominant culture. We're facing a crisis of imagination in our culture, unable to conceive of new and better strategies to organize our existence in ways that will be healthy for humans and the more-than-human world. To combat this crisis, we must work to expand and inspire our creativity, exploring imaginative alternative visions of the world.

Imagination allows students to envision the potential long-term consequences of modern social practices, to consider alternative modes of structuring society and industry that may be more sustainable, and to generate creative ideas about how to achieve those alternatives. Educational theorist Etienne Wenger states:

> Through imagination, we can locate ourselves in the world and in history, and include in our identities other meanings, other possibilities, other perspectives. It is through imagination that we recognize our own experience as reflecting broader patterns, connections, and configurations. It is through imagination that we see our own practices as continuing histories that reach far into the past, and it is through imagination that we conceive of new developments, explore alternatives, and envision possible futures.
>
> *(Wenger 1999, 178)*

This unit should help develop students' skill in imaging new and creative future possibilities, and should help them learn to apply their imagination toward productively planning and acting to create a better world. Throughout this unit, students should be stretching their expectations about what the future *could* be like. They should be learning not to accept "the way things are" as a given. This means, in part, imagining new visions of houses, cities, farms, transportation, and more. Students may envision green cities covered in urban gardens and organic buildings, or houses built into the landscape. Here one student describes ideas she took away from the activity in Lesson 8.3 in which students discuss and draw visions of the future:

Student Writing: Response Papers

by Rebecca Postowski

From the discussion we had yesterday on envisioning a future that would have sustainable practices, I really liked the idea of the magnetic ferry and electric cars. If we could cut down on the amount of pollution produced and still be able to go on our daily activities, that would be an ideal lifestyle. I also think if we could bury houses in the ground and then grow food on top of the houses, we would be able to cut down on our carbon footprint and still have the same structure that we are used to living in.

Another student also comments on remaking cities to be integrated with the natural world:

Student Writing: Response Papers

by Will Fejes

I also think that it is very important to keep nature tied into the city.

In addition to envisioning new ways to design and build things, to organize cities, and to shelter and feed ourselves, cultivating imaginative visions of better futures requires students to envision new ways to be in relationship with the natural world and with other people. It requires that students imagine living in ways that support other forms of life, nourish ecosystems, and respect the agency and participation of humans and nonhumans. It means imagining strategies that support community, connection, and positive coexistence. One anonymous student comments on the importance of working together to imagine better ways of living:

Student Writing: Response Papers

[W]e need to work in cooperative groups and strong communities so that we will have a generation of "rethinkers" that help with the ongoing process of protecting the environment and living a sustainable life.

Working together, expecting new and creative thinking, and knowing there are better possibilities – these are outcomes students should take away from this unit. I hope these lessons allow students to engage in what Marek Oziewicz (2009) describes as "social dreaming about sustainable coexistence" (85). And, by envisioning possibilities for more sustainable approaches to the world, for compassionate and positive coexistence, I hope they can begin to see ways to manifest those possibilities, and begin planning and enacting steps to achieve genuine change.

UNIT 8 READINGS AND REFERENCES

Bong, Joon-ho. 2014. *Snowpiercer*. Motion Picture. The Weinstein Company.
Callenbach, Ernest. 2004. *Ecotopia: The Notebooks and Reports of William Weston*. Berkeley, CA: Banyan Tree Books in association with Heyday Books.
Carson, Rachel. 1994. "A Fable for Tomorrow." In *Reading the Environment*, edited by Melissa Walker, 1st ed., 523–524. New York: W. W. Norton.

Emmerich, Roland. 2004. *The Day After Tomorrow*. Motion Picture. 20th Century Fox.

Fleischer, Richard. 1973. *Soylent Green*. Motion Picture. MGM.

Hemenway, Toby. 2009. *Gaia's Garden: A Guide to Home-Scale Permaculture*. 2nd ed. White River Junction, VT: Chelsea Green Publishing.

Hodge, Oliver. 2007. *Garbage Warrior*. Documentary. Open Eye Media.

Judge, Mike. 2007. *Idiocracy*. Motion Picture. 20th Century Fox.

McDonough, William, and Michael Braungart. 2002. *Cradle to Cradle: Remaking the Way We Make Things*. 1st ed. New York: North Point Press.

McKibben, Bill. 1994. "Change the Way We Think: Actions Will Follow." In *Reading the Environment*, edited by Melissa Walker, 1st ed., 567–571. New York: W. W. Norton.

Miller, George. 1982. *Mad Max 2: The Road Warrior*. Motion Picture. Kennedy Miller Productions.

Mitchell, George. 1994. "Two Children in a Future World." In *Reading the Environment*, edited by Melissa Walker, 1st ed., 525–528. New York: W. W. Norton.

Oziewicz, Marek. 2009. "'We Cooperate, or We Die': Sustainable Coexistence in Terry Pratchett's The Amazing Maurice and His Educated Rodents." *Children's Literature in Education* 40 (2): 85–94. doi:10.1007/s10583–008–9079–3.

Shiva, Vandana. 2005. *Earth Democracy: Justice, Sustainability, and Peace*. Cambridge, MA: South End Press.

Stanton, Andrew. 2008. *WALL-E*. Animated Motion Picture. Pixar Animation Studios.

Wenger, Etienne. 1999. *Communities of Practice: Learning, Meaning, and Identity*. 1st ed. New York: Cambridge University Press.

CONCLUSION

Final Thoughts on the Lessons

This book began with the question, how do we make the world what we want it to be? How do we create the more just, sustainable, and joyful world we all desire? I hope the lessons I've shared here provide the start of some strategies for doing just this. To make a better world, we must look unflinchingly at what is going on around us, think deeply about why things have come to be the way they are, ask difficult questions about our own beliefs, assumptions, and habits, build relationships of care and love with people, places, and other beings, and work together to generate new ideas that offer opportunities for joyful co-existence with the entirety of the living earth.

I have watched my students engage in each of these actions as they participate in the lessons in this book. I have watched them reflect on the tragedies of cruelty and destruction taking place around us, grieving for those who suffer and asking how we can contribute to such suffering:

Student Writing: Response Papers

by Mallory Brooks

How did we get to the point that another's life does not matter to us . . . that because an animal is not a human, they are not hurt when their family dies, or when their houses are torn down? This bothers me more than most things about the human race. We have developed such a selfish way of living that we don't even care about each other and what we do to one another none the less the animals whose homes we just tore down when we emptied a forested area to put up a housing development.

I have watched my students recognize their own role in social and environmental injustice:

Student Writing: Response Papers

by Michelle Ott

Action needs to be taken if we desire to save our planet from crumbling to pieces. We need to stop pointing fingers at everyone else and realize that we are the root of the problem. We use the products that are causing harm to the planet. We are creating the pollution. We are destroying the natural world.

I have watched them critique modern modes of living:

Student Writing: Response Papers

by Chelsie Bateman

Basically, the tree coexists with its environment whereas industrial systems do not and if that natural cycle is cut off, nothing is really growing, except for pollution and problems in nature.

I have watched them reflect on connection and disconnection:

Student Writing: Response Papers

by Heather Harshbarger

The most interesting point made is "The lines are themselves only ideas" (p 54). It's hard for me to imagine that there is only a false barrier that separates the two sides: Man from nature. It serves to remind me of the wall of light that lies between a city and the sky. When I'm at my house in a somewhat suburban neighborhood, there's no way to see the stars at night, and even the moon is a glancing chance between the buildings; but when we travel to my grandparents' house, there is not the same effect. I can sit up on the rooftop in the cool night air and actually see the stars. In the farm area, you can hear the animals and seldom any cars. It goes to show that barriers really are only something humans put up, shown by the statement by Meadows, "Even between you and me, even there, the lines are only of our own makings" (p 55).

And watched them explore different ways that we should view ourselves and the world around us:

> ## Student Writing: Response Papers
>
> *by Will Fejes*
>
> I say that we as humans, for the most part, do not think of ourselves as a part of nature. We see nature as a specific place or as the wilderness when really nature is the whole world and society, no matter the landscape or environment. We as humans are a large part of nature and our inability to see that is the main reason contributing to the downfall of our environment.

And watched them recognize that our cultural foundations have led us to this place:

> ## Student Writing: Response Papers
>
> *by Chelsie Bateman*
>
> [T]hese views will continue unless something changes in society.

My students don't always ask the same questions or reach the same conclusions, of course. Some support more dramatic change than others, but all desire some change, all want a more sustainable world. Many of them wonder whether sweeping change is possible. They question which path is right, what attitudes toward the natural world they wish to support, and what ethical standards they find reasonable and just in regard to nonhuman and human others. But they are engaged in the conversation. What's more, they feel that the conversation is necessary, and they want the knowledge and insight to approach the conversation effectively, to question, and to contribute. My students do not always agree with the course materials or with each other, but they seem excited to have the chance to explore the ideas presented in the class, glad to be able to use the sorts of language and to express the sorts of sentiments the class makes room for.

I believe, as authors like Terry Tempest Williams (1994) and Anna Peterson (2009) have suggested, that all of us want, on some level, to rediscover our innate connection, and need for connection, to the more-than-human world. We want to find ways to enact impulses of compassion, care, relationality, and love in all aspects of our lives. I hope the materials in these lessons provide a starting place for rediscovering these connections, that they offer students "glimpses of alternative values and practices" (Peterson 2009, 109), and a chance to start enacting those alternative values in their thinking and in their discussions with each other. If students can begin to see the types of changes they desire manifested in their conversations, in their thinking, in their interactions with each other and with me, then those changes may begin to seem both reasonable and possible.

At each point in these lessons, I have seen my students eager for the chance to voice fear for the future, outrage at injustice, and love for nonhuman beings, and I have seen them desperate for examples and models of ways to live in the world that are kinder, more aware, and more sustainable. Voicing these emotions, discussing the source of these problems, and remaking our perceptions of our place in the world – these are conversations that are sorely needed in our society. Providing space for such conversations to take place has been a supreme honor for me as a teacher. I hope you will have the satisfaction of experiencing conversations like those I've experienced with my students.

I hope that, in our classrooms and beyond, we can each develop more and more opportunities to voice the grief, love, and longing we feel for our world. I hope we can create more and more opportunities to reformulate our understandings, our patterns of thought, and our language. And I hope we can put these feelings, strategies, and new perspectives to good use, finding transformative paths forward built on mutual care and respect.

References

Peterson, Anna Lisa. 2009. *Everyday Ethics and Social Change: The Education of Desire*. New York: Columbia University Press.

Williams, Terry Tempest. 1994. *An Unspoken Hunger: Stories from the Field*. New York: Pantheon Books.

LIST OF READINGS

Below is a list of all the readings and films assigned in the lessons throughout this book. For each written text I have included a very brief excerpt, to give you a sense of what these readings are like. The full text of some of these documents, or links to the full text, can be found in the online resource for this book.

The excerpts provided here are included to help guide you in curriculum planning; please follow all copyright rules when acquiring and using the full text of suggested readings and films.

In the online resource for this book you'll find great examples of students writing about some of these passages, to help give you more insight into what to expect your students to get out of the materials.

Abbey, Edward. 1994. "The Serpents of Paradise." In *Reading the Environment*, edited by Melissa Walker, 1st ed., 51–56. New York: W. W. Norton.
In Lesson: 1.1
Sample: "It seems to me possible, even probable, that many of the nonhuman undomesticated animals experience emotions unknown to us. What do the coyotes mean when they yodel at the moon? . . . Precisely what did those two enraptured gopher snakes have in mind when they came gliding toward my eyes over the naked sandstone? If I had been as capable of trust as I am susceptible to fear I might have learned something new or some truth so very old we have all forgotten it" (56).

Abram, David. 1997. *The Spell of the Sensuous: Perception and Language in a More-Than-Human World.* 1st Vintage Books ed. New York: Vintage.
In Lesson: 3.1
Sample: "Apache persons often associate places with particular ancestors. Indeed, the earthly places seem to *speak* to certain persons in the voices of those grandparents who first 'shot' them with stories, or even to speak in the voices of those

long-dead ancestors whose follies and exploits are related in the *'agodzaahi* tales. The ancestral wisdom of the community resides, as it were, in the stories, but the stories – and even the ancestors themselves – reside in the land. . . . to move away from the land is ultimately to lose contact with the actual sites invoked by the place-names, and so to lose touch with the spoken stories that reside in those places" (160).

Ackerman-Leist, Philip. 2013. *Rebuilding the Foodshed: How to Create Local, Sustainable, and Secure Food Systems*. White River Junction, VT: Chelsea Green Publishing.

In Lesson: 4.2

Sample: "At the same time, local communities can work to reframe food systems that embrace a diversity of cultural and economic perspectives. Yet another approach is for self-identified groups . . . that have suffered injustice to frame their own food systems . . . some of these groups are working toward self-reliance by creating their own food systems that reflect their collective history, culture, and future aspirations" (137).

Aristotle. (350 BCE) 2014. "History of Animals." Translated by D'Arcy Wentworth Thompson. *The Internet Classics Archive*. Book IX, Part 1, paragraphs 5–7.

In Lesson: 2.4

Sample: "The fact is, the nature of man is the most rounded off and complete, and consequently in man the qualities or capacities above referred to are found in their perfection. Hence woman is more compassionate than man, more easily moved to tears, at the same time is more jealous, more querulous, more apt to scold and to strike. She is, furthermore, more prone to despondency and less hopeful than the man, more void of shame or self-respect, more false of speech, more deceptive, and of more retentive memory."

Aristotle. (350 BCE) 2014. "Politics." Translated by Benjamin Jowett. *The Internet Classics Archive*. Part XII.

In Lesson: 2.4

Sample: "A husband and father, we saw, rules over wife and children, both free, but the rule differs, the rule over his children being a royal, over his wife a constitutional rule. For although there may be exceptions to the order of nature, the male is by nature fitter for command than the female, just as the elder and full-grown is superior to the younger and more immature."

Armstrong, Virginia I. 1971. *I Have Spoken: American History Through The Voices Of The Indians*. Athens, OH: Swallow Press.

In Lesson: 6.1

Sample: "Fathers . . . it is you who are the disturbers in this land, by coming and building your towns, and taking it away unknown to us and by force. . . . The land belongs to neither one or the other, but the Great Being above allowed it to be a place of residence for us; so fathers, I desire you to withdraw" (17).

Austin, Mary. 2001. "The Land of Little Rain (1903)." In *So Glorious a Landscape: Nature and the Environment in American History and Culture*, edited by Chris J. Magoc, 92–95. Wilmington, DE: Rowman & Littlefield.

In Lesson: 6.1

Sample: "For all the toll the desert takes of a man it gives compensations, deep breaths, deep sleep, and the communion of the stars. . . . They look large and near and palpitant. . . . Wheeling to their stations in the sky, they make the poor world-fret of no account. Of no account you who lie out there watching, nor the lean coyote that stands off in the scrub from you and howls and howls" (95).

Baca, Jimmy Santiago. 2001. "Ah Rain!" In *Poetry Like Bread*, edited by Martín Espada, 54. Willimantic: Curbstone Press.

In Lesson: 1.1

Sample:

> I lift my head high,
> expand my chest to breathe! breathe! breathe!
> breath in the wood and green leaves . . .
> I drench by body, shimmer, clothes wet,
> my religion is Rain (54).

Benenson, Bill, Gene Rosow, and Eleonore Dailly. 2009. *Dirt! The Movie*. Documentary. Common Ground Media.

In Lesson: 4.3

Benton, Lisa M., and John Rennie Short, eds. 2000a. "The 1785 Ordnance." In *Environmental Discourse and Practice: A Reader*, 60–62. Malden, MA: Wiley-Blackwell.

In Lesson: 6.1

Sample: "Be it ordained by the United States in Congress assembled, that the territory ceded by individual States to the United States, which has been purchased of the Indian inhabitants, shall be disposed of in the following manner: . . . The Surveyors . . . shall proceed to divide the said territory into townships of six miles square, by lines running due north and south, and others crossing these at right angles" (60).

Benton, Lisa M., and John Rennie Short, eds. 2000b. "National Park Legislation (1864)." In *Environmental Discourse and Practice: A Reader*, 98. Malden, MA: Wiley-Blackwell.

In Lesson: 6.1

Sample: "That there shall be, and is hereby, granted to the State of California the 'cleft' or 'gorge' in the granite peak of the Sierra Nevada Mountains; known as the Yo-Semite Valley, with its branches or spurs, in estimated length fifteen miles . . . with the stipulation, nevertheless, that the said State shall accept this grant upon the express conditions that the premises shall be held for public use, resort, and recreation; shall be inalienable for all time" (89).

Benton, Lisa M., and John Rennie Short, eds. 2000c. "National Park Legislation (1872)." In *Environmental Discourse and Practice: A Reader*, 98–99. Malden, MA: Wiley-Blackwell.
In Lesson: 6.1
Sample: "[T]he tract of land in the Territories of Montana and Wyoming . . . is hereby reserved and withdrawn from settlement, occupancy, or sale under the laws of the United States, and dedicated and set apart as a public park or pleasuring-ground for the benefit and enjoyment of the people" (98–99).

Benton, Lisa M., and John Rennie Short, eds. 2000d. "National Park Legislation (1916)." In *Environmental Discourse and Practice: A Reader*, 104–105. Malden, MA: Wiley-Blackwell.
In Lesson: 6.1
Sample: "[T]here is hereby created in the Department of the Interior a service to be called the National Park Service . . . The service thus established shall promote and regulate the use of the Federal areas known as national parks, monuments, and reservations . . . which purpose is to conserve the scenery and the natural and historic objects and wild life therein" (104).

Berry, Wendell. 1995. *Another Turn of the Crank: Essays*. Washington, DC: Counterpoint.
In Lesson: 2.2, 3.1
Sample: "When we include ourselves as parts or belongings of the world we are trying to preserve, then obviously we can no longer think of the world as 'the environment' – something out there around us. We can see that our relation to the world surpasses mere connection and verges on identity. And we can see that our right to live in this world, whose parts we are, is a right that is strictly conditioned. . . . if we . . . want to live, we cannot exempt use from care. There is simply nothing in Creation that does not matter. . . . We cannot be improved – in fact, we cannot help but be damaged – by our useless or greedy or merely ignorant destruction of anything" (74–75).

Berry, Wendell. 2013. "The Pleasures of Eating." Reprinted by *The Center for Ecoliteracy*. Accessed August 29. www.ecoliteracy.org/essays/pleasures-eating.
In Lesson: 4.3
Sample: "If one gained one's whole knowledge of food from . . . advertisements (as some presumably do), one would not know that the various edibles were ever living creatures, or that they all come from the soil, or that they were produced by work. The passive American consumer, sitting down to a meal of preprepared or fast food, confronts a platter covered with inert, anonymous substances

that have been processed, dyed, breaded, sauced . . . and sanitized beyond resemblance to any part of any creature that ever lived. The products of nature and agriculture have been made, to all appearances, the products of industry. Both eater and eaten are thus in exile from biological reality. And the result is a kind of solitude, unprecedented in human experience."

Biagi, Shirley. 2011. *Media Impact: An Introduction to Mass Media*. 10th ed. Australia and Boston: Cengage Learning.
In Lesson: 2.5
Sample: "Television began as an advertising medium. Never questioning how television would be financed, the TV networks assumed they would attract commercial support. They were right. In 1949, television advertising totaled $12.3 million. In 1950, the total was $40.8 million. In 1951, advertisers spent $128 million on television. In 2010, television advertising revenue in the U.S. totaled $70 billion" (215–216).

Bigmouth, Percy. 2000. "Before They Got Thick." In *Environmental Discourse and Practice: A Reader*, edited by Lisa M. Benton and John Rennie Short, 20–21. Malden, MA: Wiley-Blackwell.
In Lesson: 6.1
Sample: "One day they looked over the big water. Then someone saw a little black dot over on the water. . . . It was a boat. The boat came to the shore. . . . All the Indians came over and watched. People were coming out. They looked at those people coming out. They saw that the people had blue eyes and were white. They thought these people might live in the water all the time" (21).

Black Elk. 2000. "The Hoop of the World." In *Environmental Discourse and Practice: A Reader*, edited by Lisa M. Benton and John Rennie Short, 257. Malden, MA: Wiley-Blackwell.
In Lesson: 2.2
Sample: "And I saw that the sacred hoop of my people was one of many hoops that made one circle, wide as daylight . . . and in the center grew one mighty flowering tree to shelter all the children of one mother and one father" (257).

Bong, Joon-ho. 2014. *Snowpiercer*. Motion Picture. The Weinstein Company.
In Lesson: 8.1

Boone, Daniel. 2000. "Moving West (1797)." In *Environmental Discourse and Practice: A Reader*, edited by Lisa M. Benton and John Rennie Short, 59–62. Malden, MA: Wiley-Blackwell.
In Lesson: 6.1
Sample: "We had passed through a great forest, on which stood myriads of trees, some gay with blossoms, others rich with fruits. Nature was here a series of wonders, and a fund of delight. Here she displayed her ingenuity and industry in

a variety of flowers and fruits, beautifully coloured, elegantly shaped, and charmingly flavoured; and we were diverted with innumerable animals presenting themselves perpetually to our view" (59).

Bowen, Kevin. 2001. "Gelatin Factory." In *Poetry Like Bread*, edited by Martín Espada, 68–69. Willimantic, CT: Curbstone Press.
In Lesson: 4.2, 7.1
Sample:
> You didn't believe at first,
> but saw the evidence:
> . . . carcasses of dead piglets . . .
> legs . . . pointed to the heavens
> that failed them (68–69).

Bozzo, Sam. 2010. *Blue Gold: World Water Wars*. Documentary. Purple Turtle Films.
In Lesson: 4.2

Bradford, William. 2000. "A Certaine Indian (1621)." In *Environmental Discourse and Practice: A Reader*, edited by Lisa M. Benton and John Rennie Short, 19–20. Malden, MA: Wiley-Blackwell.
In Lesson: 6.1
Sample: "[B]ut Squanto continued with them, and was their interpreter, and was a spetiall instrument sent of God for their good beyond their expectation. He directed them how to set their corne, wher to take fish, and to procure other comodities, and was also their pilott to bring them to unknownw places for their profitt, and never left them till he dyed" (20).

Bradford, William. 2001. "A Hideous and Desolate Wilderness (1647)." In *So Glorious a Landscape: Nature and the Environment in American History and Culture*, edited by Chris J. Magoc, 24–26. Wilmington, DE: Rowman & Littlefield.
In Lesson: 6.1
Sample: "And for the season it was winter, and they that know the winters of that country know them to be sharp and violent, and subject to cruel and fierce storms, dangerous to travel to known places, much more to search an unknown coast. Besides, what could they see but a hideous and desolate wilderness, full of wild beasts and wild men – and what multitudes there might be of them they knew not" (24–25).

Brands, H. W. 2000. "Theodore Roosevelt and Conservation." In *Environmental Discourse and Practice: A Reader*, edited by Lisa M. Benton and John Rennie Short, 113–116. Malden, MA: Wiley-Blackwell.
In Lesson: 6.1
Sample: "More than any president before him – and than most after – Roosevelt recognized that natural resources were a wasting asset if not husbanded carefully.

'We are prone to speak of the resources of this country as inexhaustible; this is not so,' he declared in his annual message at the end of 1907. Indeed, the nations' resources were being exhausted at an alarming rate" (115).

Brugnaro, Ferruccio. 2001. "Don't Tell Me Not to Bother You." In *Poetry Like Bread*, edited by Martín Espada, 75. Willimantic: Curbstone Press.
In Lesson: 1.1
Sample:

> The smoke-stacks are wounds, deep
> craters open
> on my body.
> Don't tell me to leave you alone (75).

Bullard, Robert. 2000. "Anatomy of Environmental Racism." In *Environmental Discourse and Practice: A Reader*, edited by Lisa M. Benton and John Rennie Short, 223–231. Malden, MA: Wiley-Blackwell.
In Lesson: 7.3
Sample: "People of color ... are disproportionately harmed by industrial toxins in their jobs and in their neighborhoods. These groups must contend with dirty air and drinking water – the byproducts of municipal landfills, incinerators, polluting industries, and hazardous waste treatment, storage, and disposal facilities. Why do some communities get 'dumped on' while others escape? Why are environmental regulations vigorously enforced in some communities and not in others?" (223).

Callenbach, Ernest. 2004. *Ecotopia: The Notebooks and Reports of William Weston.* Berkeley, CA: Banyan Tree Books in association with Heyday Books.
In Lesson: 8.1
Sample: "[D]own ... some other streets, creeks now run. These had earlier, at great expense, been put into huge culverts underground, as is usual in cities. The Ecotopians spent even more to bring them up to ground level again. So now on this major boulevard you may see a charming series of little falls, with water gurgling and splashing, and channels lined with rocks, trees, bamboos, ferns.... the streets are full of people.... Since practically the whole street area is 'sidewalk,' nobody worries about obstructions – or about the potholes which, as they develop in the pavement, are planted with flowers.... Ecotopians setting out to go more than a block or two usually pick up one of the sturdy white-painted bicycles that lie about the streets by the hundreds and are available free to all" (12–13).

Canty, Kristin. 2011. *Farmageddon.* Documentary. Kristin Canty Productions.
In Lesson: 4.2

Capra, Fritjof. 2002. *The Hidden Connections.* London: HarperCollins.
In Lesson: 2.3

Sample: "For a long time, scientists assumed that chimpanzee communication had nothing to do with human communication because the chimps' grunts and screams bear little resemblance to human speech. However . . . these scientists focused on the wrong channel of communication. Careful observation of chimpanzees in the wild has shown that they use their hands for much more than building tools. They are communicating with them in ways previously unimagined, gesturing to beg for food, to seek reassurance, and to offer encouragement. There are various chimpanzee gestures for 'Come with me,' 'May I pass?' and 'You are welcome,' and . . . some of these gestures differ from community to community" (56).

Carson, Rachel. 1994. "A Fable for Tomorrow." In *Reading the Environment*, edited by Melissa Walker, 1st ed., 523–524. New York: W. W. Norton.
In Lesson: 8.1
Sample: "Then a strange blight crept over the area and everything began to change. Some evil spell had settled on the community: mysterious maladies swept the flocks of chickens; the cattle and sheep sickened and died. Everywhere was a shadow of death. The farmers spoke of much illness among their families. In the town the doctors had become more and more puzzled by new kinds of sickness appearing among their patients" (523).

Carson, Rachel. 2000. "The Obligation to Endure." In *Environmental Discourse and Practice: A Reader*, edited by Lisa M. Benton and John Rennie Short, 126–128. Malden, MA: Wiley-Blackwell.
In Lesson: 2.2
Sample: "It took hundreds of millions of years to produce the life that now inhabits the earth – eons of time in which that developing and evolving and diversifying life reached a state of adjustment and balance with its surroundings. The environment, rigorously shaping and directing the life it supported, contained elements that were hostile as well as supporting. . . . Given time – time not in years but in millennia – life adjusts, and a balance has been reached. For time is the essential ingredient; but in the modern world there is not time" (126).

Chester Environmental Justice. 2008. Video recording. www.youtube.com/watch?v=5Opr-uzet7Q&feature=youtube_gdata_player.
In Lesson: 7.3

Chief Seattle. [1855] 2000. "How Can One Sell the Air?: A Manifesto for the Earth." In *Environmental Discourse and Practice: A Reader*, edited by Lisa M. Benton and John Rennie Short, 12–13. Malden, MA: Wiley-Blackwell.
In Lesson: 2.2, 6.1
Sample:

If we don't own the sweet air
and the bubbling water,

how can you buy it from us?
Each pine tree shining in the sun,
. . . is holy in the thoughts
and memory of our people (12).

Cole, Thomas. 2000. "Essay on American Scenery (1835)." In *Environmental Discourse and Practice: A Reader*, edited by Lisa M. Benton and John Rennie Short, 87–90. Malden, MA: Wiley-Blackwell.
In Lesson: 2.2, 6.1
Sample: "In the American forest we find trees in every stage of vegetable life and decay – the slender sapling rises in the shadow of the lofty tree, and the giant in his prime stands by the hoary patriarch of the wood – on the ground lie prostrate decaying ranks that once waved their verdant heads in the sun and wind. These are circumstances productive of great variety and picturesqueness – green umbrageous masses – lofty and scathed trunks – contorted branches thrust athwart the sky . . . from richer combinations than can be found in the trimmed and planted grove" (88).

Colquhoun, James, Laurentine Ten Bosch, and Carlo Ledesma. 2012. *Hungry for Change*. Documentary. Permacology Productions.
In Lesson: 4.2

Corbett, Julia. 2002. "A Faint Green Sell: Advertising and the Natural World." In *Enviropop: Studies in Environmental Rhetoric and Popular Culture*, edited by Mark Meister and Phyllis M. Japp, 141–160. Westport, CT: Praeger.
In Lesson: 2.5
Sample: "The natural world is full of cultural meaning with which to associate products, thereby attaching commodity value to qualities that are impossible to own. By borrowing and adapting well-known, stereotypical portrayals of nature, advertising is able to associate water with freshness and purity and weather as fraught with danger. If, for example, an ad wants to attach the value of 'safety' to one particular car, it might demonstrate the car's ability to dodge 'dangerous' elements of nature" (147).

Cox, Robert. 2010. *Environmental Communication and the Public Sphere*. 2nd ed. Thousand Oaks, CA: Sage Publications.
In Lesson: 2.3
Sample: "Indeed, decisions to preserve habitat for endangered species or impose regulations on factories emitting air pollution seldom result from scientific study alone. Instead, a choice to take action arises from a crucible of debate and (often) public controversy. . . . we [are] led into the realm of human communication in our study of nature and environmental problems" (26).

Cronon, William. 1995. "The Trouble with Wilderness; Or, Getting Back to the Wrong Nature." In *Uncommon Ground: Toward Reinventing Nature*, edited by William Cronon, 1st ed., 69–90. New York: W.W. Norton & Co.

In Lesson: 3.1

Sample: "For many Americans wilderness stands as the last remaining place where civilization, that all too human disease, has not fully infected the earth. It is an island in the polluted sea of urban-industrial modernity. . . . But is it? The more one knows of its peculiar history, the more one realizes that wilderness is not quite what it seems. Far from being the one place on earth that stands apart from humanity, it is quite profoundly a human creation – indeed, the creation of very particular human cultures at very particular moments in human history" (69).

Cronon, William. 2000. "Changes in the Land: Indians, Colonists, and the Ecology of New England." In *Environmental Discourse and Practice: A Reader*, edited by Lisa M. Benton and John Rennie Short, 37–44. Malden, MA: Wiley-Blackwell.

In Lesson: 6.1

Sample: "Perhaps the central contrast between Indians and Europeans at the moment they first encountered each other in New English had to do with what they saw as resources and how they thought those resources should be utilized. . . . For [colonists], perceptions of 'resources' were filtered through the language of 'commodities,' goods which could be exchanged in markets where the very act of buying and selling conferred profits on their owners. . . . Ironically, though colonists perceived fewer *resources* in New England ecosystems than did the Indians, they perceived many more *commodities*, and so committed much wider portions of those ecosystems to the marketplace" (41–42).

"The Declaration of Independence." 2010. *Ushistory.org, Independence Hall Association.* www.ushistory.org/declaration/document/index.htm.

In Lesson: 7.2

Sample: "We hold these truths to be self-evident, that all men are created equal, that they are endowed by their Creator with certain unalienable rights. . . . That to secure these rights, governments are instituted among men, deriving their just powers from the consent of the governed" (para 2).

de Nerval, Gérard. 1980. "Golden Lines." In *News of the Universe: Poems of Twofold Consciousness*, edited and translated by Robert Bly, 38. San Francisco: Sierra Club Books.

In Lesson: 1.1

Sample:

Free thinker! Do you think you are the only thinker
on this earth in which life blazes inside all things?

> Look carefully in an animal at a spirit alive;
> every flower is a soul opening out into nature . . .
> "Everything is intelligent!" And everything moves you (38).

Dillard, Annie. 1994. "Land Where the Rivers Meet." In *Reading the Environment*, edited by Melissa Walker, 1st ed., 92–93. New York: W. W. Norton.
In Lesson: 3.1
Sample: "When the shining city, too, fades, I will see only those forested mountains and hills, and the way the rivers lie flat and moving among them, and the way the low land lies wooded among them, and the blunt mountains rise in darkness from the rivers' banks, steep from the rugged south and rolling from the north" (93).

Dillard, Annie. 1994. "Living Like Weasels." In *Reading the Environment*, edited by Melissa Walker, 1st ed., 63–66. New York: W. W. Norton.
In Lesson: 1.1
Sample: "The thing is to stalk your calling in a certain skilled and supple way, to locate the most tender and live spot and plug into that pulse. This is yielding, not fighting. A weasel doesn't 'attack' anything; a weasel lives as he's meant to, yielding at every moment to the perfect freedom of single necessity" (66).

Duany, Andres, Elizabeth Plater-Zyberk, and Jeff Speck. 2000. *Suburban Nation: The Rise of Sprawl and the Decline of the American Dream*. New York: North Point Press.
In Lesson: 3.1
Sample: "Suburban sprawl, now the standard North American pattern of growth, ignores historical precedent and human experience. . . . Unlike the traditional neighborhood model, which evolved organically as a response to human needs, suburban sprawl is an idealized artificial system. It is not without a certain beauty: it is rational, consistent, and comprehensive. . . . Unfortunately, this system is already showing itself to be unsustainable" (4).

Dunayer, Joan. 2001. *Animal Equality: Language and Liberation*. Derwood, MD: Ryce.
In Lesson: 2.3, 7.1
Sample: "How do we justify our treatment of nonhumans? We lie – to ourselves and to each other, about our species and about others. Deceptive language perpetuates speciesism, the failure to accord nonhuman animals equal consideration and respect. Like sexism or racism, speciesism is a form of self-aggrandizing prejudice. . . . Speciesism can't survive without lies. Standard English usage supplies these lies in abundance. . . . Current usage promotes a false dichotomy between humans and nonhumans. Separate lexicons suggest opposite behaviors

and attributes. We eat, but other animals feed. A woman is pregnant or nurses her babies; a nonhuman mammal gestates or lactates. A dead human is a corpse, a dead nonhuman a carcass or meat" (1–2).

Durning, Alan Thein. 2006. "The Dubious Rewards of Consumption." In *The Earthscan Reader in Sustainable Consumption*, edited by Tim Jackson, 129–134. London: Earthscan.
In Lesson: 5.1
Sample: "Mutual dependence for day-to-day sustenance – a basic characteristic of life for those who have not achieved the consumer class – bonds people as proximity never can. Yet those bonds have severed with the sweeping advance of the commercial mass market into realms once dominated by family members and local enterprise. Members of the consumer class enjoy a degree of personal independence unprecedented in human history, yet hand in hand comes a decline in our attachments to each other. Informal visits between neighbours and friends, family conversation, and time spent at family meals have all diminished in the US since mid-century" (132).

"Ecuador Constitution, 2008." 2009. *Political Database of the Americas.* http://pdba.georgetown.edu/Constitutions/Ecuador/ecuador08.html.
In Lesson: 7.2
Sample: "Art. 1. Nature or Pachamama, where life is reproduced and exists, has the right to exist, persist, maintain and regenerate its vital cycles, structure, functions, and its processes in evolution. Every person, people, community . . . will be able to demand the recognition of rights for nature before the public institutions."

Emmerich, Roland. 2004. *The Day After Tomorrow.* Motion Picture. 20th Century Fox.
In Lesson: 8.1

Estabrook, Barry. 2012. *Tomatoland: How Modern Industrial Agriculture Destroyed Our Most Alluring Fruit.* Reprint ed. Andrews McMeel Publishing, LLC.
In Lesson: 4.2
Sample: "From a purely botanical and horticultural perspective, you would have to be an idiot to attempt to commercially grow tomatoes in a place like Florida. The seemingly insurmountable challenges start with the soil itself. . . . the majority of the state's tomatoes are raised in sand. Not sandy loam, not sandy soil, but pure sand, no more nutrient rich than the stuff vacationers like to wiggle their toes into on the beaches of Daytona and St. Pete. . . . In that nearly sterile medium, Florida tomato growers have to practice the equivalent of hydroponic production, only without the greenhouses" (20).

Fleischer, Richard. 1973. *Soylent Green*. Motion Picture. MGM.
In Lesson: 8.1

Foreman, Dave. 2000. "Confessions of an Eco-Warrior." In *Environmental Discourse and Practice: A Reader*, edited by Lisa M. Benton and John Rennie Short. Malden, MA: Wiley-Blackwell.
In Lesson: 2.2
Sample: "It is time for women and men, individually and in small groups, to act heroically in defense of the wild, to put a monkeywrench into the gears of the machine that is destroying natural diversity. Though illegal, this strategic monkeywrenching can be safe, easy, fun, and – most important – effective in stopping timber cutting, road building, overgrazing, oil and gas exploration, mining . . . and other forms of destruction of the wilderness, as well as cancerous suburban sprawl" (198).

Fulkerson, Lee. 2013. *Forks Over Knives*. Documentary. Monica Beach Media.
In Lesson: 4.2

Gee, James Paul. 1996. *Social Linguistics and Literacies: Ideology in Discourses*. 2nd ed. London: Taylor & Francis.
In Lesson: 2.3
Sample: "Meanings are ultimately rooted in negotiation between different social practices with different interests by people who share or seek to share some common ground. Power plays an important role in these negotiations. . . . Meaning is not something locked away in heads, rendering communication possible by the mysterious fact that everyone has the same thing in their heads. . . . It is something that has its roots in 'culture'" (12–13).

Gilpin, William. 2001. "The Untransacted Destiny of the American People (1846)." In *So Glorious a Landscape: Nature and the Environment in American History and Culture*, edited by Chris J. Magoc, 43–44. Wilmington, DE: Rowman & Littlefield.
In Lesson: 6.1
Sample: "The *untransacted* destiny of the American people is to subdue the continent – to rush over this vast field to the Pacific Ocean – to animate the many hundred millions of its people, and to cheer them upward . . . to establish a new order in human affairs – to set free the enslaved – to regenerate superannuated nations – to change darkness into light" (43).

Glenn, Cathy B. 2004. "Constructing Consumables and Consent: A Critical Analysis of Factory Farm Industry Discourse." *Journal of Communication Inquiry* 28 (1): 63–81. doi:10.1177/0196859903258573.
In Lesson: 2.3

Sample: "Most of us understand that the terms beef, veal, pork, and poultry are euphemisms employed by the industry to designate flesh from cows, calves, pigs, and birds such as chickens and turkeys. We recognize that such euphemisms are employed mainly for marketing purposes, and for the most part, we accept the practice without necessarily questioning its ethics. It is a recognizable discursive move that removes the 'beingness' or subjectivity of animals and replaces it with a word that morphs a subject for its own purposes into an object for consumption. Put differently, the use of these euphemisms disguises the fact that the body parts we purchase and consume are the objectified remains of former subjects. It is a familiar example of nature (in this case, nonhuman animals) as commodities" (70).

Gottlieb, Robert. 2005. *Forcing the Spring: The Transformation of the American Environmental Movement.* Revised ed. Washington, DC: Island Press.
In Lesson: 6.1
Sample: "The standard histories of environmentalism in the United States almost invariably begin in the West. In the vast, spectacular landscapes, in the breathtaking vistas, powerful mountain ranges, and sharp-cutting rivers, in the West's abundance and scarcity of resources, in its aridity and fertility, the forces of urbanization and industrialization created some of the most dramatic changes in environment . . . It is also in the West that much of the traditional interpretation of environmentalism is grounded" (52).

Harper, A. Breeze. 2010. *Sistah Vegan: Food, Identity, Health, and Society: Black Female Vegans Speak.* New York: Lantern Books.
In Lesson: 4.3
Sample: "Let's reflect on how our own overconsumption, unhealthy, and environmentally unsustainable patterns – indoctrinated as normal – are collectively contributing to the suffering of ourselves, nonhuman animals, and the ecosystem. We speak of how addictions to illegal drugs and alcohol can ruin entire families and neighborhoods within households and communities in the U.S. However, let's look deeper into ourselves and ask how flesh food products, cane sugar, caffeine addiction, and overconsumption in general are not only destroying our beautiful bodies, but Black and brown families, neighborhoods, and communities, locally and globally, along with the global ecosystem" (26–27).

Harrabin, Roger. 2014. "World Wildlife Populations 'Plummet.'" *BBC News.* September 30. www.bbc.com/news/science-environment-29418983.
In Lesson: 3.1
Sample: "The global loss of species is even worse than previously thought. . . . The report suggests populations have halved in 40 years. . . . humans are cutting down trees more quickly than they can re-grow, harvesting more fish than the oceans can re-stock . . . and emitting more carbon than oceans and forests can absorb" (para 1–8).

Hauter, Wenonah. 2012. *Foodopoly: The Battle Over the Future of Food and Farming in America*. Reprint ed. New York: New Press.
In Lesson: 4.1
Sample: "[T]he failure to stop massive consolidation has allowed a handful of companies to control the entire food chain – from the seeds, fertilizer, and implements to processing, distribution, and retail grocery chains. The largest twenty food companies exert tremendous control over food and farming, as both buyers of ingredients and sellers of product. . . . Reversing this corporate tyranny and concentration is critical for creating a fair and sustainable food system" (39).

Heford, Will. 2003. "That God Made." In *From Totems to Hip-Hop: A Multicultural Anthology of Poetry Across the Americas, 1900–2002*, edited by Ishmael Reed, 216–217. New York: Thunder's Mouth Press.
In Lesson: 7.3
Sample:

> These are the Timber . . . and fertile Soil
> That belong to us all in spite of the gall
> Of the Grabbers and Grafters who forestall
> The natural rights and needs of all . . . (216).

Hemenway, Toby. 2009. *Gaia's Garden: A Guide to Home-Scale Permaculture,* 2nd ed. White River Junction, VT: Chelsea Green Publishing.
In Lesson: 4.3, 8.2
Sample: "When not interrupted . . . the end result of succession nearly everywhere is forest. . . . This is why . . . suburbanites must constantly weed and chop out woody seedling from their well-watered lawns and garden beds. The typical yard, with its perfect regimen of irrigation and fertilizer, is trying hard to become a forest. . . . So why fight this trend toward woodland? Instead, we can work with nature to fashion a multistoried forest garden, a food- and habitat-producing landscape that acts like a natural woodland. In a forest garden, the yard is a parklike grove of spreading fruit trees, walnuts, chestnuts, and other useful trees. . . . Catching the sunlight farther down, dancing with birds, are flowering shrubs and berry bushes. . . . Beneath all this and in the bright edges are beds of perennial flowers, vegetables, and soil-building mulch plants" (208).

Herrera, Juan Felipe. 2003. "Earth Chorus." In *From Totems to Hip-Hop: A Multicultural Anthology of Poetry Across the Americas, 1900–2002*, edited by Ishmael Reed, 30–32. New York: Thunder's Mouth Press.
In Lesson: 2.2, 7.3
Sample:

> It is the earth that snarls and slashes with black jaguar
> eyes and teeth and incandescent claws . . .

it is the earth flowing the dew of brave women
and men on march with rivers and coffee plantations . . .
It is the earth
that triumphs (31–32).

Higman, B. W. 2011. *How Food Made History*. 1st ed. Chichester, West Sussex, UK and Malden, MA: Wiley-Blackwell.
In Lesson: 4.1
Sample: "In China, sitting on a chair to eat at a table became common only in the last thousand years. In the Shang and Chou periods, the upper ranks ate kneeling on a mat, with vessels containing the food and drink that made up their meal set out before them and arranged according to strict rules. Each man was given four bowls of grain, to fill the stomach, together with a number of meat and vegetable dishes, ranging from three to eight according to their rank and age" (150).

Hodge, Oliver. 2007. *Garbage Warrior*. Documentary. Open Eye Media.
In Lesson: 8.2

Honeyborne, James. 2013. "Elephants Really Do Grieve like Us: They Shed Tears and Even Try to 'Bury' Their Dead – a Leading Wildlife Film-Maker Reveals How the Animals Are like Us." *Mail Online*. January 30. www.dai-lymail.co.uk/news/article-2270977/Elephants-really-grieve-like-They-shed-tears-try-bury-dead-leading-wildlife-film-maker-reveals-animals-like-us.html.
In Lesson: 2.3
Sample: "Any scientist knows how dangerous it is to project human feelings on to an animal, to force them into human moulds or 'anthropomorphise' them, but it's equally dangerous to ignore a wealth of scientific data based on decades of observation in the wild. We may never know exactly what goes on inside the mind of an elephant, but it would be arrogant of us to assume we are the only species capable of feeling loss and grief" (para 3–4).

Hossay, Patrick. 2006. *Unsustainable: A Primer for Global Environmental and Social Justice*. London: Zed Books.
In Lesson: 7.3
Sample: "As the world's human population rises, the demands on the planet's natural resources increase. The . . . United Nations Environmental Program's Conservation Monitoring Center . . . has calculated changes in humanity's total demand on natural resources – the human ecological 'footprint' – for the past three decades. They have discovered that human demands on natural resources have exceeded the earth's sustainable productive capacity since the late 1970s. In fact, the human consumption of the earth's resources now exceeds the amount that the planet can sustainably provide by a full 20 per cent. . . . With individual

consumption increasing and 86 million people added to the world's population each year, this ecological deficit will grow worse fast" (34).

Houston, Pam. 1994. "A Blizzard Under Blue Sky." In *Reading the Environment*, edited by Melissa Walker, 1st ed., 57–62. New York: W. W. Norton.
In Lesson: 1.1
Sample: "One of the unfortunate things about winter camping is that it has to happen when the days are so short. Fourteen hours is a long time to lie in a snow cave. . . . I was five miles down the trail before I realized what had happened. Not once in that fourteen-hour night did I think about deadlines, or bills. . . . For the first time in many months I was happy to see a day beginning. The morning sunshine was like a present from the gods. What really happened, of course, is that I remembered about joy" (61).

Hubbell, Sue. 1994. "Mites, Moths, Bats, and Mosquitoes." In *Reading the Environment*, edited by Melissa Walker, 1st ed., 161–164. New York: W. W. Norton.
In Lesson: 2.2
Sample: "Bats are mammals like we are. They suckle their young, and have such wizened ancient-looking faces that they seem strangely akin and familiar. . . . The truth is that, from a human point of view, bats are beneficial. . . . the little brown bat and other temperate-zone bats have a diet made up . . . of night-flying insects" (162).

Jefferson, Thomas. 2011. "Notes on the State of Virginia." *Electronic Text Center, University of Virginia Library.* February 21. http://web.archive.org/web/20110221131356/http://etext.lib.virginia.edu/ etcbin/toccer-new2?id=JefVirg.sgm&images=images/modeng&data=/texts/ english/modeng/parsed&tag=public&part=14&division=div1.
In Lesson: 2.4
Sample: "But never yet could I find that a black had uttered a thought above the level of plain narration; never see even an elementary trait of painting or sculpture. In music they are more generally gifted than the whites with accurate ears for tune and time, and they have been found capable of imagining a small catch. Whether they will be equal to the composition of a more extensive run of melody, or of complicated harmony, is yet to be proved" (266).

Joanes, Ana Sofia. 2009. *Fresh*. Documentary. Ripple Effect Films.
In Lesson: 4.2

Jordan, Chris. 2005. "Intolerable Beauty: Portraits of American Mass Consumption." *Chris Jordan Photographic Arts.* www.chrisjordan.com/gallery/ intolerable/#cellphones2.
In Lesson: 5.1

Jordan, Chris. 2011. "Midway: Message from the Gyre." *Chris Jordan Photographic Arts.* www.chrisjordan.com/gallery/midway/#CF000313%2018x24.
In Lesson: 5.1

Judge, Mike. 2007. *Idiocracy.* Motion Picture. 20th Century Fox.
In Lesson: 8.1

Kennedy, Scott Hamilton. 2014. *The Garden.* Documentary. Black Valley Films.
In Lesson: 4.2

Kenner, Robert. 2008. *Food, Inc.* Documentary. Magnolia Pictures.
In Lesson: 4.2

Kingsolver, Barbara. 2007. *Animal, Vegetable, Miracle: A Year of Food Life.* New York: HarperCollins.
In Lesson: 4.3
Sample: "We wanted to live in a place that could feed us: where rain falls, crops grow, and drinking water bubbles right up out of the ground. This might seem an abstract reason for leaving beloved friends and one of the most idyllic destination cities in the United States. But it was real to us" (3).

Klein, Naomi. 2010. "A Hole in the World." *The Nation.* June 24. www.thenation.com/article/36608/hole-world?page=0,0.
In Lesson: 2.2
Sample: "In the arc of human history, the notion that nature is a machine for us to re-engineer at will is a relatively recent conceit. . . . Europeans – like indigenous people the world over – believed the planet to be a living organism. . . . The metaphor changed with the unlocking of some (but by no means all) of nature's mysteries during the Scientific Revolution of the 1600s. With nature now cast as a machine, devoid of mystery or divinity, its component parts could be dammed, extracted and remade with impunity" (3).

Koons, Deborah. 2004. *The Future of Food.* Documentary. Lily Films.
In Lesson: 4.2

Kunstler, James Howard. 1994. *The Geography of Nowhere: The Rise and Decline of America's Man-Made Landscape.* 1st Touchstone ed. New York: Simon & Schuster.
In Lesson: 3.1
Sample: "There was nothing like it before in history: a machine that promised liberation from the daily bondage of place. And in a free country like the United States, with the unrestricted right to travel, a vast geographical territory to spread out into, and a national tradition of picking up and moving whenever life at home became intolerable, the automobile came as a blessing. In the early years of

motoring, hardly anyone understood the automobile's potential for devastation – not just of the landscape or air, but of culture in general" (86).

Lakoff, George, and Mark Johnson. 1980. *Metaphors We Live By*. Chicago: University of Chicago Press.
In Lesson: 2.2
Sample: "Our ordinary conceptual system, in terms of which we both think and act, is fundamentally metaphorical in nature. . . . the way we think, what we experience, and what we do every day is very much a matter of metaphor. . . . To give some idea . . . let us start with . . . the conceptual metaphor ARGUMENT IS WAR. . . . we don't just *talk* about arguments in terms of war. . . . We see the person we are arguing with as an opponent. We attack his positions and we defend our own. . . . The very systematicity that allows us to comprehend one aspect of a concept in terms of another (e.g., comprehending an aspect of arguing in terms of battle) will necessarily hide other aspects of the concept. In allowing us to focus on one aspect of a concept . . . a metaphorical concept can keep us from focusing on other aspects . . . that are inconsistent with that metaphor" (3–10).

Lappé, Anna, and Bryant Terry. 2006. *Grub: Ideas for an Urban Organic Kitchen*. New York: Tarcher.
In Lesson: 4.1
Sample: "More than 35 million Americans live in food-insecure households in which they are often unsure where their next meal is coming from. . . . The United States has lost roughly a third of our farms since the early 70s. And many of the surviving farmers are no longer independent, but are contract workers for large food companies. . . . Four pounds of pesticides for every American man, woman, and child are used every year in the United States, more than one-fifth of the world total for all pesticides. . . . Yet crop loss to pests for some crops is double what it was before this chemical storm. . . . The typical American diet is now implicated in diseases ranging from hypertension to certain types of cancer to Type 2 diabetes" (4–5).

Leopold, Aldo. 1970. *A Sand County Almanac: With Essays on Conservation from Round River*. 1st Ballantine Books ed. New York: Ballantine Books.
In Lesson: 1.1, 2.2, 2.3, 3.1, 7.1 (These lessons contain different passages by Leopold.)
Sample: "There are two spiritual dangers in not owning a farm. One is the danger of supposing that breakfast comes from the grocery, and the other that heat comes from the furnace. To avoid the first danger, one should plant a garden. . . . To avoid the second, he should lay a split of good oak on the andirons . . . and let it warm his shins while a February blizzard tosses the trees outside. If one has cut, split, hauled, and piled his own good oak . . . he will remember much about

where the heat comes from. . . . The particular oak now aglow on my andirons grew on the bank of the old emigrant road where it climbs the sandhill. . . . It shows 80 growth rings, hence the seedling from which it originated must have laid its first ring of wood in 1865. . . . It was a bolt of lightning that put an end to wood-making by this particular oak. We mourned the loss of the old tree, but knew that a dozen of its progeny standing straight and stalwart on the sands had already taken over its job of wood-making. We let the dead veteran season for a year in the sun it could no longer use, and then on a crisp winter's day we laid a newly filed saw to its bastioned base" (6–9).

Levenstein, Harvey A. 1993. *Paradox of Plenty: A Social History of Eating in Modern America.* New York: Oxford University Press.
In Lesson: 4.1
Sample: "Family meals were on the wane, even where traditional family structures survived. Snacking, on the other hand, was becoming a continuous process, indulged in practically at all times in all places. Indeed, it was calculated that three-quarters of all Americans derived at least 20 percent of their energy needs from snacks" (236).

Lim, Genny. 2003. "Animal Liberation." In *From Totems to Hip-Hop: A Multicultural Anthology of Poetry Across the Americas, 1900–2002*, edited by Ishmael Reed, 34–36. New York: Thunder's Mouth Press.
In Lesson: 1.1
Sample:

[A]s soon as her body makes contact
with liquid
There is instant recognition
She dives into the pool and emerges with her feathers wet and glistening . . .
My heart sings to see this once captive duck
Frolic in the lake, diving and dancing, flapping her wings (35–36).

Lorca, Federico García. 1980. "New York (Office and Attack)." In *News of the Universe: Poems of Twofold Consciousness*, edited and translated by Robert Bly, 110–112. San Francisco: Sierra Club Books.
In Lesson: 1.1, 7.1
Sample:

The . . . hogs and the lambs
lay their drops of blood down
underneath all the statistics;
and the terrible bawling of the packed-in cattle
fills the valley with suffering
where the Hudson is getting drunk on its oil (111).

Lyon, George Ella. 1996. "Where I'm From." In *United States of Poetry*, edited by Joshua Blum, Bob Holman, and Mark Pellington, 22–23. New York: Harry N. Abrams.
In Lesson: 3.1
Sample:

> I am from the dirt under the back porch. . . .
> I am from . . . the Dutch elm
> whose long gone limbs I remember
> as if they were my own (22).

Magoc, Chris J., ed. 2001a. "Acoma Pueblo Creation Myth." In *So Glorious a Landscape: Nature and the Environment in American History and Culture*, 20–22. Wilmington, DE: Rowman & Littlefield.
In Lesson: 6.1
Sample: "In the beginning two female human beings were born. These two children were born underground at a place called Shipapu. As they grew up, they began to be aware of each other. There was no light and they could only feel each other. Being in the dark, they grew slowly" (20).

Magoc, Chris J., ed. 2001b. "Tewa Sky Looms." In *So Glorious a Landscape: Nature and the Environment in American History and Culture*, 23. Wilmington, DE: Rowman & Littlefield.
In Lesson: 6.1
Sample:

> Oh our Mother the Earth, oh our Father the Sky,
> Your children are we, and with tired backs
> We bring you the gifts that you love (23).

Manno, Jack. 2002. "Commoditization: Consumption Efficiency and an Economy of Care and Connection." In *Confronting Consumption*, edited by Thomas Princen, Michael F. Maniates, and Ken Conca, 1st ed., 67–99. Cambridge, MA: MIT Press.
In Lesson: 5.1
Sample: "Under the current industrial capitalist system of incentives and disincentives, what we consider 'progress' is invariably directed toward increasing levels of consumption. There are, however, other possible paths of progress and development. . . . technical progress that improves the efficiency of labor could be directed toward either increasing production or reducing labor time. If directed toward freeing up more and more time, this time could be directed toward improvement in community and cultural life. This time would become the raw material for an economy of care and connection" (70).

Marsh, George Perkins. 2001. "The Destructiveness of Man (1864)." In *So Glorious a Landscape: Nature and the Environment in American History and Culture*, edited by Chris J. Magoc, 136–139. Wilmington, DE: Rowman & Littlefield.

In Lesson: 6.1
Sample: "But man is everywhere a disturbing agent. Wherever he plants his foot, the harmonies of nature are turned to discords. . . . Indigenous vegetable and animal species are extirpated, and supplanted by others of foreign origin, spontaneous production is forbidden or restricted, and the face of the earth is either laid bare or covered with a new and reluctant growth of vegetable forms, and with alien tribes of animal life" (137).

Marshall, Andrew Gavin. 2012. "No Conspiracy Theory – A Small Group of Companies Have Enormous Power Over the World." *AlterNet.* October 31. www.alternet.org/world/no-conspiracy-theory-small-group-companies-have-enormous-power-over-world.
In Lesson: 5.1
Sample: "This 'core' was found to own roughly 80% of global revenues for the entire set of 43,000 TNCs. And then came what the researchers referred to as the 'super-entity' of 147 tightly-knit companies, which all own each other, and collectively own 40% of the total wealth in the entire network" (paragraph 2).

Martusewicz, Rebecca A., Jeff Edmundson, and John Lupinacci. 2011. *EcoJustice Education: Toward Diverse, Democratic, and Sustainable Communities.* 1st ed. New York: Routledge.
In Lesson: 2.2, 2.4
Sample: "Discourses use root metaphors to structure and maintain a 'that's the way it is' perception of the world. When these discursive patterns are shared and exchanged by large groups of people, they create a complex world-view – that is, a deeply ingrained set of ideas that structures how one sees, relates to and behaves in the world" (66).

McDonough, William, and Michael Braungart. 2002. *Cradle to Cradle: Remaking the Way We Make Things.* 1st ed. New York: North Point Press.
In Lesson: 5.1, 8.2
Sample: "Consider the cherry tree: thousands of blossoms create fruit for birds, humans, and other animals, in order that one pit might eventually fall onto the ground, take root, and grow. Who would look at the ground littered with cherry blossoms and complain, 'How inefficient and wasteful!' The tree makes copious blossoms and fruit without depleting its environment. Once they fall on the ground, their materials decompose and break down into nutrients that nourish microorganisms, insects, plants, animals, and soil. Although the tree actually makes more of its 'product' than it needs for its own success in an ecosystem, this abundance has evolved . . . to serve rich and varied purposes. In fact, the tree's fecundity nourishes just about everything around it. What might the human-built world look like if a cherry tree had produced it?" (72–73).

McKibben, Bill. 1990. *The End of Nature*. 1st Anchor Books ed. New York: Anchor Books.
In Lesson: 2.2
Sample: "By the end of nature I do not mean the end of the world. The rain will still fall and the sun shine, though differently than before. When I say 'nature,' I mean a certain set of human ideas about the world and our place in it. But the death of those ideas begins with concrete changes in the reality around us – changes that scientists can measure and enumerate. . . . these changes will clash with our perceptions, until, finally, our sense of nature as eternal and separate is washed away" (8).

McKibben, Bill. 1993. *The Age of Missing Information*. New York: Plume.
In Lesson: 2.5
Sample: "We believe that we live in the 'age of information,' that there has been an information 'explosion,' an information 'revolution.' While in a certain narrow sense this is the case, in many important ways just the opposite is true. We also live at a moment of deep ignorance, when vital knowledge that humans have always possessed about who we are and where we live seems beyond our reach. An Unenlightenment. An age of missing information. This account of that age takes the form of an experiment – a contrast between two days. One day, May 3, 1990, lasted well more than a thousand hours – I collected on videotape nearly every minute of television that came across the enormous Fairfax cable system from one morning to the next, and then I watched it all. The other day, later that summer, lasted the conventional twenty-four hours. A mile from my house, camped on a mountaintop by a small pond, I awoke, took a day hike up a neighboring peak, returned to the pond for a swim, made supper, and watched the stars till I fell asleep. This book is about the results of that experiment – about the information that each day imparted" (9–10).

McKibben, Bill. 1994. "Change the Way We Think: Actions Will Follow." In *Reading the Environment*, edited by Melissa Walker, 1st ed., 567–571. New York: W. W. Norton.
In Lesson: 8.2
Sample: "The difficulty is almost certainly more psychological than intellectual – less that we can't figure out major alterations in our way of life than that we simply don't want to. Even if our way of life has destroyed nature and endangered the planet, it is so hard to imagine living in any other fashion" (571).

McKibben, Bill. 2008. *Deep Economy: The Wealth of Communities and the Durable Future*. Later printing. New York: St. Martin's Griffin.
In Lesson: 5.1
Sample: "A single-minded focus on increasing wealth has driven the planet's ecological systems to the brink of failure, without making us happier. . . . economists built us a wonderful set of tools for getting More. And those tools work. . . . It's

easy to understand why they, and the political leaders they advise, would be pleased to try and keep using those tools – pleased to keep us . . . achieving ever greater economies of scale. But there's something profoundly unrealistic and sentimental about that approach" (42–45).

Meadows, Donella. 1994. "What Is Biodiversity and Why Should We Care about It?" In *Reading the Environment*, edited by Melissa Walker, 1st ed., 149–151. New York: W. W. Norton.
In Lesson: 2.2
Sample: "Suppose you were assigned to turn every bit of dead organic matter from fallen leaves to urban garbage to road kills into nutrients that feed new life. Even if you knew how, what would it cost? Uncountable numbers of bacteria, molds, mites, and worms do it for free. If they ever stopped, all life would stop. We would not last long. If green plants stopped turning our exhaled carbon dioxide back into oxygen. The plants would not last long if a few beneficent kinds of soil bacteria stopped turning nitrogen from the air into fertilizer" (149).

Meadows, Donella. 1999. "Lines in the Mind." In *Our Land, Ourselves: Readings on People and Place*, edited by Peter Forbes, Ann Armbrecht, and Helen Whybrow, 2nd ed., 53–55. San Francisco: Trust for Public Land.
In Lesson: 2.2
Sample: "The earth was formed whole and continuous in the universe, without lines. The human mind arose in the universe needing lines, boundaries, distinctions. . . . Between me and not-me there is surely a line, a clear distinction, or so it seems. But now that I look, where is that line? . . . When I drink, the waters of the earth become me. With every breath I take in I draw in not-me and make it me. . . . If the air and the waters and the soils are poisoned, I am poisoned. Only if I believe the fiction of the lines more than the truth of the lineless planet will I poison the earth, which is myself " (53–54).

"Media/Political Bias." 2014. *Rhetorica*. Accessed October 28. http://rhetorica.net/bias.htm.
In Lesson: 2.5
Sample: "The news media are money-making businesses. As such, they must deliver a good product to their customers to make a profit. The customers of the news media are advertisers. The most important product the news media delivers to its customers are readers or viewers. Good is defined in numbers and quality of readers or viewers. The news media are biased toward conflict (re: bad news and narrative biases below) because conflict draws readers and viewers" (para 8).

Merchant, Carolyn. 2000. "Ecofeminism." In *Environmental Discourse and Practice: A Reader*, edited by Lisa M. Benton and John Rennie Short, 209–213. Malden, MA: Wiley-Blackwell.

In Lesson: 2.2
Sample: "Ecofeminism emerged in the 1970s with an increasing consciousness of the connections between women and nature. . . . an ecological revolution would entail new gender relations between women and men and between
humans and nature" (209).

Merkel, Jim. 2003. *Radical Simplicity: Small Footprints on a Finite Earth*. Gabriola
Island, BC: New Society Publishers.
In Lesson: 5.1
Sample: "Loving our limits can set the stage for our life. As we recognize that
we only have one Earth – which has finite capacity to support life – becoming
comfortable with limits will open our minds and hearts for the work of taming
the appetite. . . . there are infinite satisfying lifestyle packages compatible with
living on a finite, equitable share of nature" (16).

Midway Film and Chris Jordan. 2012. *Midway Trailer*. Video Recording. www.
midwayfilm.com/index.html.
In Lesson: 5.1

Miller, George. 1982. *Mad Max 2: The Road Warrior*. Motion Picture. Kennedy
Miller Productions.
In Lesson: 8.1

Miller, Joaquin. 2001. "Social and Environmental Degradation in the California
Gold Country (1890)." In *So Glorious a Landscape: Nature and the Environment
in American History and Culture*, edited by Chris J. Magoc, 46–49. Wilmington,
DE: Rowman & Littlefield.
In Lesson: 6.1
Sample: "But now the natives of these forests. . . . They do not smite the mountain rocks for gold, nor fell the pines, nor roil up the waters and ruin them for
the fishermen. All this magnificent forest is their estate. The Great Spirit made
this mountain first of all, and gave it to them, they say, and they have possessed
it ever since. They preserve the forest, keep out the fires, for it is the park for
their deer" (47).

Mitchell, George. 1994. "Two Children in a Future World." In *Reading the
Environment*, edited by Melissa Walker, 1st ed., 525–528. New York: W. W.
Norton.
In Lesson: 8.1
Sample: "Journey with me for a moment into the twenty-first century and meet
two children, Luisa and Eric. Luisa lives in Mexico City sometime in the next
century, about fifty years from now. She looks up at a polluted sky in this future
world and sees no sun at all. In all her young life – she is now nine – she has rarely

seen the sun. She has heard that Mexico City was once a beautiful place to live. There was plenty of sun once, and a lot more room" (525).

Molla, Rani. 2014. "Can Organic Farming Counteract Carbon Emissions?" *Wall Street Journal: The Numbers.* May 22. http://blogs.wsj.com/numbers/can-organic-farming-counteract-carbon-emissions-1373/.
In Lesson: 4.3
Sample: "Organic practices could counteract the world's yearly carbon dioxide output while producing the same amount of food as conventional farming, a new study suggests. . . . if all cropland were converted to the regenerative model it would sequester 40% of annual CO_2 emissions; changing global pastures to that model would add another 71%, effectively overcompensating for the world's yearly carbon dioxide emissions" (para 1–3).

Momaday, N. Scott. 1999. "The Man Made of Words." In *Our Land, Ourselves: Readings on People and Place*, edited by Peter Forbes, Ann Armbrecht, and Helen Whybrow, 2nd ed., 71–73. San Francisco: Trust for Public Land.
In Lesson: 1.1
Sample: "None of us lives apart from the land entirely; such an isolation is unimaginable. We have sooner or later to come to terms with the world around us. . . . We Americans need now more than ever before – and indeed more than we know – to imagine who and what we are with respect to the earth and sky" (72).

Monson, Shaun. 2005. *Earthlings*. Documentary. Nation Earth.
In Lesson: 4.2

Monterey Californian. 2001. "Americans Spread All Over California (1846)." In *So Glorious a Landscape: Nature and the Environment in American History and Culture*, edited by Chris J. Magoc, 45. Wilmington, DE: Rowman & Littlefield.
In Lesson: 6.1
Sample: "[T]here can be little doubt, that the industry, and intelligence of the agriculturists, which are daily emigrating to this country, will improve the nature of the soil to such a degree, as to greatly augment both the produce, and improve the flavour of this most delicious fruit, and the same may be said of all the other production of this country" (45).

Mühlhäusler, Peter. 2003. *Language of Environment, Environment of Language: A Course in Ecolinguistics*. London: Battlebridge.
In Lesson: 2.3
Sample: "Markedness is the term that is increasingly common in ecolinguistic writing. . . . What is grammatically less marked is acquired earlier. It is regarded as normal in a particular society and is statistically more frequent. The fact that *he*

rather than *she* has been the unmarked third person pronoun in English is seen to reflect the fact that sexism has been the normal state of affairs in English society. The fact that the verb 'to see' in its unmarked sense stands for human vision (rather than insect vision) and the unmarked meaning of 'to walk' is 'human movement' yet another example of anthropocentrism in language. . . . Seeing is portrayed as the unmarked form of any sensory perception and the absence of visual perception tends to be negatively spoken of as blindness: bats, moles and worms are called blind which suggests that they have some major handicap" (22).

Muir, John. 2000. "A Voice for Wilderness (1901)." In *Environmental Discourse and Practice: A Reader*, edited by Lisa M. Benton and John Rennie Short, 102–104. Malden, MA: Wiley-Blackwell.
In Lesson: 2.2, 6.1
Sample: "That any one would try to destroy [Hetch Hetchy Valley] seems incredible. . . . These temple destroyers, devotees of ravaging commercialism, seem to have a perfect contempt for Nature, and, instead of lifting their eyes to the God of the mountains, lift them to the Almighty Dollar. Dam Hetch Hetchy! As well dam for water-tanks the people's cathedrals and churches, for no holier temple has ever been consecrated" (104).

Muir, John. 2001. "My First Summer in the Sierra (1868)." In *So Glorious a Landscape: Nature and the Environment in American History and Culture*, edited by Chris J. Magoc, 80–83. Wilmington, DE: Rowman & Littlefield.
In Lesson: 6.1
Sample: "Never before had I seen so glorious a landscape, so boundless an affluence of sublime mountain beauty. The most extravagant description I might give of this view to any one who has not seen similar landscapes with his own eyes would not so much as hint its grandeur and the spiritual glow that covered it" (81).

National Association Opposed to Woman Suffrage. 2014. "Vote NO on Woman Suffrage." Pamphlet reprinted in The Atlantic.com. Accessed October 11. www.theatlantic.com/sexes/archive/2012/11/vote-no-on-womens-suffrage-bizarre-reasons-for-not-letting-women-vote/264639/.
In Lesson: 2.4
Sample: "Because in some States more voting women than voting men will place the Government under petticoat rule. Because it is unwise to risk the good we already have for the evil which may occur."

Nixon, Richard. 2000. "Message to Congress." In *Environmental Discourse and Practice: A Reader*, edited by Lisa M. Benton and John Rennie Short, 132–139. Malden, MA: Wiley-Blackwell.
In Lesson: 2.2
Sample: "The recent upsurge of public concern over environmental questions reflects a belated recognition that man has been too cavalier in his relations with

nature. Unless we arrest the depredations that have been inflicted so carelessly on our natural systems—which exist in an intricate set of balances—we face the prospect of ecological disaster" (132).

Orr, David. 2010. "The Carbon Connection." *Center for Ecoliteracy*. Accessed August 29. www.ecoliteracy.org/essays/carbon-connection.
In Lesson: 5.1
Sample: "Like all life forms, we search out great pools of carbon to perpetuate ourselves. It is our mismanagement of carbon that threatens the human future. . . . The exploitation of carbon is the original sin, leading quite possibly to the heat death of a great portion of life on Earth, including us" (para 20).

Oziewicz, Marek. 2009. "'We Cooperate, or We Die': Sustainable Coexistence in Terry Pratchett's The Amazing Maurice and His Educated Rodents." *Children's Literature in Education* 40 (2): 85–94. doi:10.1007/s10583-008-9079-3.
In Lesson: 8.2
Sample: "I shall argue that The Amazing Maurice develops the theme of sustainable coexistence. . . . The 'transformative purpose' . . . of the book seems obvious. A story about rats and a cat who develop consciousness and try to make it in a world dominated by humans; a story at the climax of which rat characters are presented as more humane than humans; a story which ends with an unprecedented pact, negotiated by a cat, between rats and humans who establish a shared civilization; such elements clearly suggest to young and not-so-young readers that the ideas of cooperation and coexistence are being pushed to their limit" (87).

Parasecoli, Fabio. 2008. *Bite Me: Food in Popular Culture*. New York: Berg.
In Lesson: 4.1
Sample: "Food is pervasive. The social, economic, and even political relevance cannot be ignored. Ingestion and incorporation constitute a fundamental component of our connection with reality and the world outside our body. Food influences our lives as a relevant marker of power, cultural capital, class, gender, ethnic, and religious identities. . . . Yet, food reveals many other layers of meaning that are often left unexplored when it comes to phenomena that fall squarely under the heading of pop culture" (2).

Parkin, Katherine. 2000. "Campbell's Soup and the Long Shelf Life of Traditional Gender Roles." In *Kitchen Culture in America: Popular Representations of Food, Gender, and Race*, edited by Sherrie A. Inness, 51–64. Philadelphia: University of Pennsylvania Press.
In Lesson: 4.1
Sample: "Domestic reformer Catharine Beecher, feminist theorist Charlotte Perkins Gilman, and a host of their contemporaries in the nineteenth century advocated changes in cooking practices, including increasing its efficiency and moving it out of the home altogether. With minimal fanfare, food advertising

co-opted these once radical ideas and helped deradicalize them into normal aspects of daily life in the twentieth century. Yet the transformation of cooking and eating patterns was not the result of an organized movement to liberate women from kitchen drudgery. While it may have lessened their work, canned foods did not challenge women's exclusive responsibility for the homemaking role" (53).

Parkman, Francis, Jr. 2000. "The Oregon Trail (1849)." In *Environmental Discourse and Practice: A Reader*, edited by Lisa M. Benton and John Rennie Short, 62–63. Malden, MA: Wiley-Blackwell.
In Lesson: 6.1
Sample: "It was a remarkably fresh and beautiful May morning. The rich and luxuriant woods through which the miserable road conducted us, were lighted by the bright sunshine and enlivened by a multitude of birds. We overtook on the way our late fellow-travellers, the Kanzas Indians, who, adorned with all their finery, were proceeding homeward at a round pace" (63).

Patel, Raj. 2008. *Stuffed and Starved: The Hidden Battle for the World Food System*. Brooklyn, NY: Melville House.
In Lesson: 4.1
Sample: "International trade transformed the world and, in its high capitalist form, was premised on a great deal of exploitation, for a wide range of goods, across large parts of the planet. Slave labour was an integral part of the provision of cheap food to European cities. African slaves were an essential component, for instance, of the plantation economies of the United States, Caribbean, and Brazil" (81).

Peterson, Anna Lisa. 2009. *Everyday Ethics and Social Change: The Education of Desire*. New York: Columbia University Press.
In Lesson: 7.1
Sample: "We find the presence and beginning of hope for a better world in loving friendships, family ties, encounters with nonhuman creatures, and ventures into wild nature. These relationships and experiences give meaning and value to our lives. They find us at our best: relating to human and nonhuman others in nonutilitarian ways, sacrificing for larger goods, finding satisfaction in experience and relationship rather than consumption and calculation. Because they embody not only alternative values but also alternative sources of joy, these experiences constitute the immanent utopias of everyday life" (3).

Pinchot, Gifford. 2000. "The Birth of Conservation." In *Environmental Discourse and Practice: A Reader*, edited by Lisa M. Benton and John Rennie Short, 116–118. Malden, MA: Wiley-Blackwell.
In Lesson: 6.1

Sample: "It had never occurred to us that we were all parts of one another . . . the fact that the Federal Government had taken up the protection of the various natural resources individually and at intervals during more than half a century doubtless confirmed our bureaucratic nationalism. . . . Instead of being, as we should have been, like a squadron of cavalry, all acting together for a single purpose, we were like loose horses in a field, each one following his own nose" (117).

Pollan, Michael. 2007. *The Omnivore's Dilemma: A Natural History of Four Meals.* New York: Penguin.
In Lesson: 4.1
Sample: "You are what you eat, it's often said, and if this is true, then what we mostly are is corn – or, more precisely, processed corn. . . . researchers who have compared the isotopes in the flesh or hair of North Americans to those in the same tissues of Mexicans report that it is now we in the North who are the true people of the corn. . . . So that's us: processed corn, walking" (20–23).

Princen, Thomas. 2002. "Distancing: Consumption and the Severing of Feedback." In *Confronting Consumption*, edited by Thomas Princen, Michael F. Maniates, and Ken Conca, 1st ed., 103–131. Cambridge, MA: MIT Press.
In Lesson: 5.1
Sample: "The propensity to externalize costs through production processes that separate production and consumption decisions. . . . Divorce the primary production decisions . . . from the selling and consumption decisions. One result is that, as key decisions are removed from primary producers and costs are externalized, feedback from the resource (e.g., land, fishery, forest) . . . is severed. . . . A third result is that as distance increases along any dimension and, as a result, feedback is severed, restraint in resource consumption diminishes. . . . Even the most committed environmental altruist and the broadest thinking global citizen cannot know of, or have influence on, production and selling decisions at a distance" (126–127).

Quintana, Leroy V. 2001. "Sharks." In *Poetry Like Bread*, edited by Martín Espada, 202–203. Willimantic: Curbstone Press.
In Lesson: 1.1
Sample:
> We swim towards blood the way
> some people cross an imaginary line
> into the United States to pick lettuce . . .
> Like the lettuce pickers and dishwashers,
> all we are good for is fighting, trouble (202–203).

Riedelsheimer, Thomas. 2002. *Rivers and Tides: Andy Goldsworthy Working with Time.* Documentary. Roxie Releasing.
In Lesson: 1.1

Rifkin, Jeremy. 1994. "Big, Bad Beef." In *Reading the Environment*, edited by Melissa Walker, 1st ed., 20–21. New York: W. W. Norton.
In Lesson: 4.2
Sample: "The beef addiction of the U.S. and other industrialized nations has also contributed to the global food crisis. Cattle and other livestock consume more than 70 percent of the grain produced in the U.S. and about a third of the world's total grain harvest. . . . If the U.S. land now used to grow livestock feed were converted to grow grain for human consumption, we could feed an additional 400 million people" (20).

Rilke, Rainer Maria. 1980. "I Live My Life." In *News of the Universe: Poems of Twofold Consciousness*, edited by Robert Bly, 76. San Francisco: Sierra Club Books.
In Lesson: 1.1
Sample:
> And I still don't know if I am a falcon,
> Or a storm, or a great song (76).

Rilke, Rainer Maria. 1995. "Ah, Not to Be Cut off." In *Ahead of All Parting: The Selected Poetry and Prose of Rainer Maria Rilke*, 191. New York: Modern Library.
In Lesson: 1.1
Sample:
> not through the slightest partition
> shut out from the law of the stars (191).

Robin, Marie-Monique. 2008. *The World According to Monsanto*. Documentary. Image et Compagnie.
In Lesson: 4.2

Roosevelt, Theodore. 2000. "Conservation, Protection, Reclamation, and Irrigation (1901)." In *Environmental Discourse and Practice: A Reader*, edited by Lisa M. Benton and John Rennie Short. Malden, MA: Wiley-Blackwell.
In Lesson: 2.2, 6.1
Sample: "Wise forest protection does not mean the withdrawal of forest resources, whether of wood, water, or grass, from contributing their full share to the welfare of the people, but, on the contrary, gives the assurance of larger and more certain supplies. The fundamental idea of forestry is the perpetuation of forests by use. Forest protection is not an end of itself; it is a means to increase and sustain the resources of our country and the industries which depend upon them" (111).

Root, Waverly, and Richard De Rochemont. 1981. *Eating In America*. Hopewell, NJ: Ecco.
In Lesson: 4.1
Sample: "It was an Indian habit to stow away caches of long-lasting foods in various places where they might one day be needed; it was the Pilgrims' good

luck to stumble on one of these caches, which kept them alive (some of them) over their first terrible winter. In the South, the Indians furnished the newcomers directly with enough food to assure survival (for some of them). Later the Indians would become less inclined to promote the survival of Europeans" (54).

Ryan, John C., and Alan Thein Durning. 1997. *Stuff: The Secret Lives of Everyday Things*. Seattle, WA: Northwest Environment Watch.
In Lesson: 4.1, 5.1
Sample: "Workers earning less than a dollar a day picked my coffee berries by hand and fed them into a diesel-powered crusher, which removed the beans from the pulpy berries that encased them. The pulp was dumped into the Cauca River. The beans, dried under the sun, traveled to New Orleans on a ship in a 132-pound bag. For each pound of beans, about two pounds of pulp had been dumped into the river. As the pulp decomposed, it consumed oxygen needed by fish in the river. . . . At New Orleans. . . . the beans were packaged in four-layer bags constructed of poly-ethylene, nylon, aluminum foil, and polyester. They were trucked to a Seattle warehouse in an 18-wheeler, which got six miles per gallon of diesel" (9–10).

Sanford, J. B. 2014. "Argument Against Women's Suffrage, 1911." San Francisco Public Library. Accessed October 11. http://sfpl.org/pdf/libraries/main/sfhistory/suffrageagainst.pdf.
In Lesson: 2.4
Sample: "The men are able to run the government and take care of the women. Do women have to vote in order to receive the protection of man? Why, men have gone to war, endured every privation and death itself in defense of woman. . . . By keeping woman in her exalted position man can be induced to do more for her than he could by having her mix up in affairs that will cause him to lose respect and regard for her."

Sears, John F. 1999. *Sacred Places: American Tourist Attractions in the Nineteenth Century*. Reprint ed. Amherst: University of Massachusetts Press.
In Lesson: 6.1
Sample: "Visitors to the Falls sometimes referred to themselves as 'pilgrims.' . . . [Tourists] struggled over slippery rocks, were wetted with spray and buffeted by winds. Such tourist rituals recapitulate, if only faintly, the trials experienced by pilgrims on their journey toward Canterbury, Rome, or Jerusalem and their initiation into the mysteries of the sacred place" (13).

Shakur, Tupac. 1999. "The Rose That Grew from Concrete." In *The Rose That Grew from Concrete*, 3. New York: MTV/Pocket Books.
In Lesson: 1.1
Sample:
>Long live the rose that grew from concrete
>when no one else ever cared! (3)

Shiva, Vandana. 2000. *Stolen Harvest: The Hijacking of the Global Food Supply.* Cambridge, MA: South End Press.
In Lesson: 4.2
Sample: "[C]orporate control of food and globalization of agriculture are robbing millions of their livelihoods and their right to food. . . . It is being experienced in every society, as small farms and small farmers are pushed to extinction, as monocultures replace biodiverse crops, as farming is transformed from the production of nourishing and diverse foods into the creation of markets for genetically engineered seeds, herbicides, and pesticides. . . . the myth of 'free trade' and the global economy becomes a means for the rich to rob the poor of their right to food and even their right to life. For the vast majority of the world's people – 70 percent – earn their livelihoods by producing food. The majority of these farmers are women" (7).

Shiva, Vandana. 2005. *Earth Democracy: Justice, Sustainability, and Peace.* Cambridge, MA: South End Press.
In Lesson: 7.2, 8.2
Sample: "All beings are subjects who have integrity, intelligence, and identity, not objects of ownership, manipulation, exploitation, or disposability. No humans have the right to own other species, other people, or the knowledge of other cultures through patents and other intellectual property rights" (9).

Siebert, Charles. 2014. "Should a Chimp Be Able to Sue Its Owner?" *New York Times Magazine.* April 23. www.nytimes.com/2014/04/27/magazine/the-rights-of-man-and-beast.html.
In Lesson: 7.2
Sample: "Yet while animal-welfare laws . . . statutes now abound, the primary thrust of such legislation remains the regulation of our various uses and abuses of animals, including food production, medical research, entertainment and private ownership. The fundamental legal status of nonhumans, however, as things, as property, with no rights of their own, has remained unchanged. Wise has devoted himself to subverting that hierarchy by moving the animal from the defendant's table to the plaintiff's" (para 20–21).

Silko, Leslie Marmon. 1986. *Ceremony.* New York: Penguin Books.
In Lesson: 3.1
Sample: "He believed then that . . . on certain nights, when the moon rose full and wide as a corner of the sky, a person standing on the high sandstone cliff of that mesa could reach the moon. Distances and days existed in themselves then; they all had a story. They were not barriers. If a person wanted to get to the moon, there was a way; it all depended on whether you knew the directions – exactly which way to go and what to do to get there; it depended on whether you knew the story of how others before you had gone. He had believed in

the stories for a long time, until the teachers at Indian school taught him not to believe in that kind of 'nonsense'" (19).

Singer, Peter. 2002. *Animal Liberation*. New York: Ecco.
In Lesson: 7.1
Sample: "It is an implication of this principle of equality that our concern for others and our readiness to consider their interests ought not to depend on what they are like or on what abilities they may possess. Precisely what our concern or consideration requires us to do may vary according to the characteristics of those affected by what we do: concern for the well-being of children growing up in America would require that we teach them to read; concern for the well-being of pigs may require no more than that we leave them with other pigs in a place where there is adequate food and room to run freely. But the basic element – the taking into account of the interests of the being, whatever those interests may be – must, according to the principle of equality, be extended to all beings, black or white, masculine or feminine, human or nonhuman. . . . there are no good reasons, scientific or philosophical, for denying that animals feel pain. . . . Animals can feel pain. As we saw earlier, there can be no moral justification for regarding the pain (or pleasure) that animals feel as less important than the same amount of pain (or pleasure) felt by humans" (5–15).

Singer, Peter, and Jim Mason. 2006. *The Way We Eat: Why Our Food Choices Matter*. Emmaus, PA: Rodale.
In Lesson: 4.2
Sample: "In 2005, the world began to face the serious possibility that the cheap chicken produced by factory farming could be far more costly to all of us than even the most radical animal rights advocates had ever dreamt it might be. Scientists began to warn leaders to prepare against the possibility of an epidemic of avian influenza – popularly known as bird flu . . . Supporters of factory farming have used the threat of bird flu to make a case against having chickens outdoors, claiming that the virus can be spread by migrating birds to free-range flocks. But the real danger, as scientists now recognize, is intensive chicken production" (34).

Sire, James W. 1988. *The Universe Next Door: A Basic Worldview Catalog*. 2nd ed. Downers Grove, IL: InterVarsity Press.
In Lesson: 2.1
Sample: "The very first thing every one of us recognizes before we even begin to think at all is that something exists. In other words, all world views assume that something is there rather than that nothing is there. This assumption is so primary most of us don't even know we are assuming it. . . . What we discover quickly, however, is that once we have recognized *that* something is there, we have not necessarily recognized *what* that something is. And here is where world views begin to diverge. Some people assume (with or without thinking about it) that

the only basic substance that exists is *matter*. For them, everything is ultimately one thing. Others agree that everything is ultimately one thing, but assume that one thing is Spirit or Soul or some such nonmaterial substance" (17).

Snyder, Gary. 1994a. "The Call of the Wild." In *Reading the Environment*, edited by Melissa Walker, 1st ed., 71–74. New York: W. W. Norton.
In Lesson: 1.1
Sample:

> So they bomb and they bomb
> Day after day, across the planet
> blinding sparrows . . .
> splintering trunks of cherries . . .
> Dumping poisons and explosives . . .
> A war against earth (72).

Snyder, Gary. 1994b. "The World Is Places." In *Reading the Environment*, edited by Melissa Walker, 1st ed., 88–91. New York: W. W. Norton.
In Lesson: 3.1
Sample: "Our place is part of what we are. Yet even a 'place' has a kind of fluidity: it passes through space and time. . . . A place will have been grasslands, then conifers, then beech and elm. It will have been half riverbed, it will have been scratched and plowed by ice. And then it will be cultivated, paved, sprayed, dammed, graded, built up. . . . The whole earth is a great tablet holding the multiple overlaid new and ancient traces of the swirl of forces" (90).

Snyder, S. 2014. "The Great Chain of Being." *Grand View University Faculty Pages.* Accessed October 28. http://faculty.grandview.edu/ssnyder/121/121%20great%20chain.htm.
In Lesson: 2.4
Sample: "[E]ach link in the Great Chain of Being represented a distinct category of living creature. . . . Those creatures or things higher on the Chain possessed greater intellect, movement, and ability than those placed below. . . . Humans in turn were believed to possess greater attributes than animals. Thus it was proper for them to rule over the rest of the natural world" (para 3).

Soussan, Tania. 2004. "Scientist: Prairie Dogs Have Own Language." *redOrbit.* December 4. www.redorbit.com/news/display/?id=108412.
In Lesson: 2.3
Sample: "Prairie dogs . . . are talking up a storm. They have different 'words' for tall human in yellow shirt, short human in green shirt, coyote, deer, red-tailed hawk and many other creatures. They can even coin new terms for things they've never seen before . . . according to Con Slobodchikoff . . . biology professor and prairie dog linguist" (para 1–2).

Sproul, Barbara C. 1991. *Primal Myths: Creation Myths Around the World.* 1st HarperCollins ed. San Francisco: HarperSanFrancisco.
In Lesson: 2.1, 6.1
Sample: "When God came to the earth to prepare the present order of things, he found three beings there, the thunder, an elephant, and a Dorobo, all living together. . . . The thunder declared that he was afraid of the man and said he would run away and go to the heavens. The man seeing the thunder go away was pleased, and said: 'The person I was afraid of has fled. . . . ' He then went to the woods and made some poison into which he dipped an arrow, and . . . and shot the elephant. . . . The elephant died, and the man became great in all the countries" (48–49).

Stanton, Andrew. 2008. *WALL-E.* Animated Motion Picture. Pixar Animation Studios.
In Lesson: 8.1

Steinbeck, John. 1997. *The Grapes of Wrath.* New York: Penguin Books.
In Lesson: 6.1
Sample: "In the roads where the teams moved, where the wheels milled the ground and the hooves of the horses beat the ground, the dirt crust broke and the dust formed. Every moving thing lifted the dust into the air: a walking man lifted a thin layer as high as his waist, and a wagon lifted the dust as high as the fence tops, and an automobile boiled a cloud behind it" (5).

Stone, Christopher D. 2010. *Should Trees Have Standing? Law, Morality, and the Environment.* 3rd ed. New York: Oxford University Press.
In Lesson: 7.1
Sample: "It is not inevitable, nor is it wise, that natural objects should have no rights to seek redress in their own behalf. It is no answer to say that streams and forests cannot have standing because streams and forests cannot speak. Corporations cannot speak, either; nor can states, estates, infants, incompetents, municipalities, or universities. Lawyers speak for them, as they customarily do for the ordinary citizen with legal problems. One ought, I think, to handle the legal problems of natural objects as one does the problems of legal incompetents – human beings who have become vegetative. If a human being shows signs of becoming senile and has affairs that he is de jure incompetent to manage, those concerned with his well being make such a showing to the court, and someone is designated by the court with the authority to manage the incompetent's affairs. The guardian (or 'conservator' or 'committee' – the terminology varies) then represents the incompetent in his legal affairs. Courts make similar appointments when a corporation has become 'incompetent': they appoint a trustee in bankruptcy or reorganization to oversee its affairs and speak for it in court when that becomes necessary" (8).

Story of Stuff Project. 2007. *The Story of Stuff*, with Annie Leonard. Video recording. www.youtube.com/watch?v=9GorqroigqM&feature=youtube_gdata_player.
In Lesson: 5.1

Suzuki, David T, and Wayne Grady. 2004. *Tree: A Life Story*. Vancouver: Greystone Books.
In Lesson: 3.1
Sample: "Trees are communal, sometimes to the point of being communistic: they grow together in large groups, as though for comfort or protection. They have relationships – including sexual relationships through cross-pollination – and even communicate with other trees within their stands, including tress of their own kind as well as those of other species; they function for the benefit of the whole in sometimes startling ways; and they enter into mutualistic partnerships with other species – even other species so distantly related they belong to different orders – as surely as human beings raise beans for food" (45–46).

Svendsen, Gro. 2000. "Letters Home (1863–1865)." In *Environmental Discourse and Practice: A Reader*, edited by Lisa M. Benton and John Rennie Short, 64–66. Malden, MA: Wiley-Blackwell.
In Lesson: 6.1
Sample: "I have often thought that I ought to tell you about life here in the New world. Everything is so totally different from what it was in our beloved Norway. You never will really know what it's like, although you no doubt try to imagine what it might be. Your pictures would be all wrong, just as mine were" (64).

Swenson, May. 2003. "Weather." In *From Totems to Hip-Hop: A Multicultural Anthology of Poetry Across the Americas, 1900–2002*, edited by Ishmael Reed, 52–53. New York: Thunder's Mouth Press.
In Lesson: 2.2
Sample:
> I hope they never get a rope on you, weather. . . .
> Reteach us terror, weather,
> With your teeth on our ships,
> your hoofs on our houses (52–53).

Szymborska, Wisława. 1998a. "The Silence of Plants." In *Poems: New and Collected 1957–1997*, 269–270. San Diego: Harcourt.
In Lesson: 1.1
Sample:
> I'll explain as best I can, just ask me:
> what seeing with two eyes is like,
> what my heart beats for,
> and why my body isn't rooted down (270).

Szymborska, Wisława. 1998b. "Among the Multitudes." In *Poems: New and Collected 1957–1997*, 267–268. San Diego: Harcourt.

In Lesson: 1.1

Sample:

> I could have been someone . . . much less fortunate,
> bred for my fur
> or Christmas dinner . . .
> A tree rooted to the ground
> as the fire draws near (267).

Szymborska, Wisława. 1998c. "Water." In *Poems: New and Collected 1957–1997*, 58–59. San Diego: Harcourt.

In Lesson: 2.2

Sample:

> A drop of water fell on my hand,
> drawn from the Ganges and the Nile. . . .
> Someone was drowning, someone dying was
> calling out for you (58).

Tapahonso, Luci. 1997. *Blue Horses Rush In: Poems and Stories*. Tucson: University of Arizona Press.

In Lesson: 3.1

Sample:

> The sky is a blanket of stars covering all of us.
> The night is folding darkness girl. . . .
> In the dark stillness, slants of moon and starlight
> wait within the curved walls for white dawn girl (5–6).

Thoreau, Henry David. 1994. "Walking." In *Reading the Environment*, edited by Melissa Walker, 1st ed., 41–44. New York: W. W. Norton.

In Lesson: 1.1

Sample: "Hope and the future for me are not in lawns and cultivated fields, not in towns and cities, but in the impervious and quaking swamps. I derive more of my subsistence from the swamps which surround my native town than from the cultivated gardens in the village" (43).

Thoreau, Henry David. 2001. "Where I Lived and What I Lived For (1854)." In *So Glorious a Landscape: Nature and the Environment in American History and Culture*, edited by Chris J. Magoc, 74–79. Wilmington, DE: Rowman & Littlefield.

In Lesson: 6.1

Sample: "Let us spend one day as deliberately as Nature, and not be thrown off the track by every nutshell and mosquito's wing that falls on the rails. Let us rise early and fast, or break fast, gently and without perturbation; let company come and let company go, let the bells ring and the children cry – determined to make a day of it" (78).

Turner, Frederick Jackson. 2000. "The Significance of the Frontier in American History (1894)." In *Environmental Discourse and Practice: A Reader*, edited by Lisa M. Benton and John Rennie Short, 75–77. Malden, MA: Wiley-Blackwell.
In Lesson: 6.1
Sample: "But the most important effect of the frontier has been in the promotion of democracy here and in Europe. As has been indicated, the frontier is productive of individualism. Complex society is precipitated by the wilderness into a kind of primitive organization based on the family" (76).

Turner, Rita. 2010. "Discourses of Consumption in US-American Culture." *Sustainability* 2 (7): 2279–2301.
In Lesson: 5.1
Sample: "Turn on most television networks or radio stations, or open most magazines, and you'll likely find a plethora of aspiration narratives extolling the glamour, excitement, beauty, and happiness that arise from and can be derived from material acquisition. Such messages are part of one of the most long-standing and highly-visible discourses about consumption in US-American culture: the discourse surrounding the idea that 'more is better'" (2284).

United Nations Office of the High Commissioner for Human Rights. 2011. "Eco-Farming Can Double Food Production in 10 Years, Says New UN Report." *United Nations Human Rights*. March 8. www.ohchr.org/EN/NewsEvents/Pages/DisplayNews.aspx?NewsID=10819&LangID=E.
In Lesson: 4.3
Sample: "Small-scale farmers can double food production within 10 years in critical regions by using ecological methods. . . . Agroecology applies ecological science to the design of agricultural systems that can help put an end to food crises and address climate-change and poverty challenges" (para 1–3).

"U.S. Constitution." 2010. *Legal Information Institute*. Accessed August 31. http://topics.law.cornell.edu/constitution.
In Lesson: 7.2
Sample: "The President shall. . . . from time to time give the Congress information of the state of the union, and recommend to their consideration such measures as he shall judge necessary. . . . he shall receive ambassadors and other public ministers, he shall take care that the laws be faithfully executed, and shall commission all the officers of the United States."

Viegas, Jennifer. 2005. "Chickens Worry About the Future." *ABC Science*. July 15. www.abc.net.au/science/articles/2005/07/15/1415178.htm.

In Lesson: 2.3
Sample: "Chickens don't just live in the present, but can anticipate the future and demonstrate self-control, something previously attributed only to humans and other primates, according to a recent study" (para 1).

Wackernagel, Mathis, and Williams E. Rees. 1996. *Our Ecological Footprint: Reducing Human Impact on the Earth (New Catalyst Bioregional Series).* Gabriola Island, BC: New Society Publishers.
In Lesson: 5.1
Sample: "Ecological footprint analysis is an accounting tool that enables us to estimate the resource consumption and waste assimilation requirements of a defined human population or economy in terms of a corresponding productive land area. Typical questions we can ask with this tool include: how dependent is our . . . population on resource imports from 'elswhere' and on the waste assimilation capacity of the global commons? . . . By revealing how much land is required to support any specified lifestyle indefinitely, the Ecological Footprint concept demonstrates the continuing material dependence of human beings on nature" (9–11).

Walker, Alice. 1994. "The Place Where I Was Born." In *Reading the Environment,* edited by Melissa Walker, 1st ed., 94–98. New York: W. W. Norton.
In Lesson: 3.1
Sample: "I cried one day as I talked to a friend about a tree I loved as a child. . . . During my childhood . . . I looked up at it frequently and felt reassured by its age, its generosity despite its years of brutalization . . . and its tall, old-growth pine nobility. When it was struck by lightning and killed, and cut down and made into firewood, I grieved as it had been a friend. Secretly. Because who among the members of my family would not have laughed at my grief?" (95)

Walker, Alice. 2014. "Am I Blue." *The Westcoast Post.* Accessed October 28. http://westcoastword.wordpress.com/2013/06/01/am-i-blue-by-alice-walker/.
In Lesson: 2.4
Sample: "Occasionally, when he came up for apples, he looked at me. It was a look so piercing, so full of grief, a look so human, I almost laughed (I felt too sad to cry) to think there are people who do not know that animals suffer. People like me who have forgotten, and daily forget, all that animals try to tell us."

Warren, Karen. 2000. *Ecofeminist Philosophy: A Western Perspective on What It Is and Why It Matters.* Lanham, MD: Rowman & Littlefield.
In Lesson: 2.4, 7.1
Sample: "Euro-American language is riddled with examples of 'sexist-naturist language,' that is, language that depicts women, animals, and nonhuman nature as inferior to . . . men and male-identified culture. . . . Animalizing women *in a*

patriarchal culture where animals are seen as inferior to humans, thereby reinforces and authorizes women's inferior status. Similarly, language that feminizes nature *in a patriarchal culture*, where women are viewed as subordinate and inferior, reinforces and authorizes the domination of nature" (27).

White, Lynn. 1967. "The Historical Roots of Our Ecologic Crisis." *Science* 155 (3767): 1203–1207.
In Lesson: 2.1
Sample: "[V]iewed historically . . . modern technology is at least partly to be explained as a . . . realization of the Christian dogma of man's transcendence of, and rightful mastery over, nature. . . . What we do about ecology depends on our ideas of the man–nature relationship. More science and more technology are not going to get us out of the present ecologic crisis until we find a new religion, or rethink our old one" (155).

Williams, Terry Tempest. 1994. *An Unspoken Hunger: Stories from the Field.* New York: Pantheon Books.
In Lesson: 1.1, 3.1 (These lessons contain a few different passages by Williams.)
Sample: "If I choose not to become attached to nouns – a person, place, or thing – then when I refuse an intimate's love or hoard my spirit, when a known landscape is bought, sold, and developed, chained or grazed to a stubble, or a hawk is shot and hung by its feet on a barbed-wire fence, my heart cannot be broken because I never risked giving it away. . . . Our lack of intimacy with each other is in direct proportion to our lack of intimacy with the land. We have taken our love inside and abandoned the wild. . . . The land is love. Love is what we fear. To disengage from the earth is our own oppression" (64–65).

Wilson, E. O. 1994. "Storm over the Amazon." In *Reading the Environment,* edited by Melissa Walker, 1st ed., 151–160. New York: W. W. Norton.
In Lesson: 2.2
Sample: "Biological diversity . . . is the key to the maintenance of the world as we know it. Life in a local site struck down by a passing storm springs back quickly because enough diversity still exists. . . . species evolved for just such an occasion rush in to fill the spaces. . . . This is the assembly of life that took a billion years to evolve. It has . . . created the world that created us. It holds the world steady" (160).

Winne, Mark. 2009. *Closing the Food Gap: Resetting the Table in the Land of Plenty.* Reprint ed. Boston: Beacon Press.
In Lesson: 4.2
Sample: "The reemergence of hunger at this level, as a national phenomenon, stunned everybody. But in addition to the response it evoked, it also brought into focus ironies in the nation's food system and the gap between the haves and the have-nots. In the same year, 1982, that soup lines were longer than at anytime

since the Great Depression, the federal government was stockpiling surplus cheese from dairy farmers faster than it could give it away. Grain harvests were so abundant that farmers were going out of business due to record low prices" (27–28).

Woolf, Aaron. 2007. *King Corn*. Documentary. ITVS.
In Lesson: 4.2

Wright, Ronald. 2005. *A Short History of Progress*. New York: Da Capo Press.
In Lesson: 2.1, 5.1
Sample: "The myth of progress has sometimes served us well – those of us seated at the best tables, anyway. . . . But I shall argue in this book that it has also become dangerous. Progress has an internal logic that can lead beyond reason to catastrophe. A seductive trail of successes may end in a trap. . . . Technology is addictive. Material progress creates problems that are – or seem to be – soluble only by further progress. Again, the devil here is in the scale: a good bang can be useful; a better bang can end the world" (5–8).

Wulf, Andrea. 2012. *Founding Gardeners: The Revolutionary Generation, Nature, and the Shaping of the American Nation*. Reprint ed. New York: Vintage.
In Lesson: 6.1
Sample: "The founding fathers' passion for nature, plants, gardens and agriculture is woven deeply into the fabric of America and aligned with their political thought, both reflecting and influencing it. In fact, I believe, it's impossible to understand the making of America without looking at the founding fathers as farmers and gardeners" (4).

Young, Al. 2003. "Seeing Red." In *From Totems to Hip-Hop: A Multicultural Anthology of Poetry Across the Americas, 1900–2002*, edited by Ishmael Reed, 69. New York: Thunder's Mouth Press.
In Lesson: 4.2
Sample:

> Now . . . you actually poison me through irradiation.
> Where's your imagination? Where's the spirit
> of the Aztecs, who grew me to death, named me tamatl,
> and loved me for the very fruity berry that I am? (69)

Zwinger, Ann. 1994. "The Lake Rock." In *Reading the Environment*, edited by Melissa Walker, 1st ed., 100–105. New York: W. W. Norton.
In Lesson: 3.1
Sample: "Encircling the rock is the community of plants and animals which can survive only in the water. Small motes of existence, they float with its currents, cling to underwater supports, or burrow in the brown silt of the lake bottom. . . . I watch a fat trout lurking in the fringed shadows of the sedges. All around the edges of the lake, where water meets land, grow willows, sedges and rushes" (100).

INDEX